# Spehlmann's
# Evoked Potential Primer

Second Edition

# Spehlmann's
# Evoked Potential Primer

## Visual, Auditory, and Somatosensory
## Evoked Potentials in Clinical Diagnosis

Revised and updated by

KARL E. MISULIS, M.D., Ph.D.
*Associate Clinical Professor of Neurology
Vanderbilt University, Nashville, Tennessee
and Neurologist, Semmes-Murphey Clinic
Jackson, Tennessee*

**Butterworth–Heinemann**
Boston   London   Oxford   Singapore   Sydney   Toronto   Wellington

**Library of Congress Cataloging-in-Publication Data**

Misulis, Karl E.
    Spehlmann's evoked potential primer : visual, auditory, and
somatosensory evoked potentials in clinical diagnosis. — Revised
and updated / by Karl E. Misulis.
        p.    cm.
    Rev. ed. of: Evoked potential primer / Rainer Spehlmann, c1985.
    Includes bibliographical references and index.
    ISBN 0-7506-9512-9
    1. Evoked potentials (Electrophysiology)  2. Nervous system—
Diseases—Diagnosis.  I. Spehlmann, Rainer,  Evoked
potential primer.  II. Title.  III. Title: Evoked potential primer.
    [DNLM:  1. Evoked Potentials.  WL 102 M6785 1994]
RC386.6.E86S63  1994
616.8'047547--dc20
DNLM/DLC
for Library of Congress                                    94-6178
                                                            CIP

British Library Cataloguing-in-Publication Data

A catalogue record for this book is available from the British Library.

Butterworth–Heinemann
313 Washington Street
Newton, MA 02158

10 9 8 7 6 5 4 3 2 1

Printed in the United States of America

# Contents

**Part B**
**Visual Evoked Potentials**

# Preface

Rainer Spehlmann, M.D., was an excellent neurophysiologist who wrote two classic books, *EEG Primer* and *Evoked Potential Primer*. These continue to be widely read. His untimely death prevented us from enjoying updates to these marvelous texts. Luckily, his heirs have allowed us to use his work as a foundation for new texts in each of these subjects. We hope that Dr. Spehlmann's wisdom and educational expertise will continue to inspire generations of new clinical neurophysiologists.

Evoked potentials continue to be an important clinical tool for the neurologist and neurosurgeon, though their use has changed in recent years. This is not only because of advances in evoked potential technology, but also because of improvements in imaging techniques. Evoked potentials are still used for diagnosis of demyelinating disease and spinal cord lesions, but their clinical utility for intracerebral lesions has waned. Intraoperative evoked potential technology has exploded; its use is now routine in many neurosurgical practices.

This text introduces the reader to evoked potentials. For each modality, we present the theory, methods, and interpretation.

# List of Acronyms

| | |
|---|---|
| AC | alternating current |
| A/D | analog-to-digital (conversion) |
| AEP | auditory evoked potential |
| AMA | American Medical Association |
| BAEP | brainstem auditory evoked potential |
| c | cycle |
| cd | candela (unit of light brightness) |
| CFPD | critical frequency of photic driving |
| CM | cochlear microphonic |
| CMCT | central motor conduction time |
| CNV | contingent negative variation |
| CT | computerized tomography |
| dB | decibels |
| DC | direct current |
| deg | degree (of arc) |
| ECochG | electrocochleogram |
| EEG | electroencephalogram |
| EKG | electrocardiogram |
| EMG | electromyogram |
| EP | evoked potential |
| ERA | electric response audiometry |
| ERG | electroretinogram |
| ERP | event-related potential |
| FAT | file allocation table |
| FFP | frequency following potential |
| HFF | high-frequency filter |
| HL | hearing level |
| Hz | Hertz (measure of frequency; 1 cycle per second) |
| IPL | interpeak latency |
| LED | light-emitting diode |

| | |
|---|---|
| LFF | low-frequency filter |
| LLAEP | long-latency auditory evoked potential |
| $m^2$ | meter (m) = square meter) |
| MEP | motor evoked potential |
| MLAEP | middle-latency auditory evoked potential |
| MRI | magnetic resonance imaging |
| MS | multiple sclerosis |
| NAP | nerve action potential (usually implies auditory nerve) |
| nHL | normal hearing level |
| peSPL | peak sound pressure level |
| PRVEP | pattern reversal VEP |
| PSVEP | pattern shift VEP |
| sec | second |
| SEP | somatosensory evoked potential |
| SL | sensation level |
| SNAP | sensory nerve action potential |
| SP | summating potential |
| SPL | sound pressure level |
| VEP | visual evoked potential |
| ° | degree; if unspecified indicates degrees of arc |
| ′ | minutes of arc |
| ″ | seconds of arc |
| $e$ | base of natural logarithms; approximately 2.7 |
| π | pi; approximately 3.14 |
| μ | micro, one millionth |

# A

# Evoked Potentials

---

## PART CONTENTS

---

# 1

# General Description of Evoked Potentials

## 1.1 DEFINITION

*Evoked potentials* (EPs) are the electric responses of the nervous system to motor or sensory stimulation. Classically, the stimuli have been sensory, but motor evoked potentials may become an important clinical tool. The potentials consist of a sequence of waves, each of which has a specific latency, amplitude, and polarity. The stimuli are delivered by electrical stimulation of the skin, visual stimuli, auditory stimuli, or stimulation of the motor cortex. The recordings are made by surface electrodes over the limbs, spinal cord, or brain. Needle electrodes and depth electrodes are sometimes used in research; they are not used in routine clinical practice.

A single response to a stimulus is of low amplitude and frequently obscured by electrical activity unrelated to the stimulus. By averaging many stimulus–response trials, the evoked potential rises out of the background. Electrical activity unrelated to the stimulus is averaged out of the recording. The term *evoked potential* is defined as the average of multiple responses. The term *evoked response* is defined as the elec-

trical recording following a single stimulus. The term *peak*, or *wave*, is defined as the positive or negative deflections that make up the EP. The term *component* is defined as an individual contribution to the potential, such as *low-frequency component* or *late component*.

The term *event-related potential* (ERP) is defined as a potential related to cognitive or initiative processes. These are not widely used in clinical practice, so are only briefly discussed in this book.

## 1.2 CLINICAL USE OF EPS

EPs test conduction in visual, auditory, somatosensory, and motor systems. They are most sensitive for detection of lesions in the spinal cord and brain, and much less useful for detection of peripheral nerve lesions. Historically, the main use of EPs was in detection of clinically silent lesions in suspected multiple sclerosis (MS), spinal cord dysfunction, and optic nerve lesions. New advances in imaging techniques have made EPs less useful for some of these, especially MS,

TABLE 1.1. EP types
_____

A. Stimulus modality and type
  1. Visual evoked potentials (VEPs)
    a. VEP to checkerboard pattern stimulation
    b. VEP to diffuse light stimulation
    c. VEPs to other stimuli
  2. Auditory evoked potentials (AEPs)
    a. Short-latency AEP
    b. Middle-latency AEP
    c. Long-latency AEP
    d. Other AEPs
  3. Somatosensory evoked potentials
    a. SEP to arm nerve stimulation
    b. SEP to leg nerve stimulation
    c. Other SEPs
  4. Motor evoked potentials
B. EP origin
  1. Cortical EP
  2. Subcortical EP
    a. Brainstem EP
    b. Spinal EP
    c. Brachial plexus EP
    d. Cauda equina EP
    e. Sensory nerve action potential (SNAP)
C. Recording site
  1. Scalp EP
  2. Neck EP
  3. Clavicular EP
  4. Lumbosacral EP
D. Recording method
  1. Near-field EP
  2. Far-field EP
E. Stimulus rate
  1. Transient EP
  2. Stready-state EP
F. Stimulus duration
  1. EP to brief stimuli
  2. EP to onset or end of long stimuli
  3. EP to gradually changing stimuli
G. Unilateral and bilateral stimuli
  1. EP to unilateral stimulation
  2. EP to bilateral stimulation
H. Midline and lateral recordings
  1. EP recorded in the midline
  2. EPs recorded unilaterally
    a. EP ipsilateral to the stimulus
    b. EP contralateral to the stimulus
  3. EPs recorded bilaterally

but they will continue to be used. EPs differ from magnetic resonance imaging (MRI) and computerized tomography (CT) in that they rely on integrity of the membrane and molecular systems responsible for axonal transport and synaptic transmission.

Intraoperative monitoring will probably become the predominant use of EPs. Spinal surgery relies on somatosensory evoked potentials (SEPs) to alert the surgeon to the possibility of cord traction. SEPs are also used occasionally to monitor cerebral function during carotid endarterectomy. Auditory evoked potentials (AEPs) are often used during posterior fossa surgery. Recently, AEPs have been used for evaluation of brainstem function in neonates, and in patients receiving audiotoxic chemotherapy. VEPs are often monitored during surgery of the orbit.

## 1.3 TYPES OF EPS

EPs are distinguished on the basis of stimulus and recording methods (Table 1.1). For each major modality, there are subdivisions, discussed in the respective individual sections of this book.

### 1.3.1 Stimulus Modality and Type

The individual stimulus modalities are visual evoked potentials (VEPs), auditory evoked potentials (AEPs), somatosensory evoked potentials (SEPs), and motor evoked potentials (MEPs). VEPs are subdivided into the type of visual stimulus: flash, checkerboard, and other specific patterns of stimuli. SEPs are divided by location of stimulus, with the most common being median, peroneal, and tibial nerves.

All EP modalities can be divided into short, middle, and long latency. The short-latency responses are clinically most important for AEPs and SEPs. These are the most reproducible, and least susceptible to drugs and volitional influences.

### 1.3.2 Generator of the EP

EPs are generated by the brain or spinal cord, depending on the stimulus mode. The generators are either synaptic transmission or the charge movement due to propagating action potentials in a nerve tract. The exact location of the generator will be discussed for each modality.

### 1.3.2.1 Cortical EPs

Cortical EPs are generated by synaptic transmission in the cortex and/or by the movement of charge in thalamocortical projections. SEPs and VEPs employ cortical responses in routine interpretation. Damage to the cortex or of projections to the cortex will interfere with these potentials.

AEPs used for clinical interpretation do not include cortical potentials. Late components to the responses may be generated in cortical projections; however, these are variable and sensitive to environmental factors, and therefore not used in clinical interpretation.

### 1.3.2.2 Subcortical EPs

Subcortical EPs refer strictly to potentials generated by subcortical nuclei and projections to the cortex. There is possible overlap with cortical EPs, since we now know that the cortical potential may have a contribution from these subcortical potentials. Potentials generated in the brainstem and spinal cord are clearly subcortical.

## 1.3.3 Recording Site

Location of recording electrodes determines the recorded potentials. SEPs use separate electrodes for recording from the spinal cord and brain. If the active and reference electrodes are both over the cortex, they record only the cortical and immediate subcortical potential. In contrast, if an active electrode is over the cortex and the reference electrode is over a distant part of the body, both subcortical and cortical potentials will be recorded.

## 1.3.4 Recording Method

The concept of near-field and far-field potentials is important for understanding the generators of the EPs. To a certain extent the terms are self-explanatory, but the electrophysiological implications are tremendous.[56]

### 1.3.4.1 Near-field recording

Near-field recording means that the neuronal potential passes immediately below the recording electrode. The electrode picks up the direct flow of charge between the area of depolarization and the area of repolarization. The amplitude of a near-field potential can be large, although cortical near-field potentials are attenuated by scalp, skull, and other cranial tissues.

The near-field recording shows a brief initial positive wave followed by a larger negative wave, which in turn is followed by another positive wave. The initial positivity is due to electrotonic depolarization of the membrane under the electrode. The subsequent negativity is due to the wave of depolarization passing underneath the electrode. The final positivity is due to the wave of repolarization. The major depolarization is a negative potential even though an intracellular recording would show a reduction in the negative membrane potential. This is because the depolarization is due to the influx of positively charged sodium and calcium ions. The movement of these cations into the cells produces a negative potential with extracellular recording.

### 1.3.4.2 Far-field recording

Far-field potentials are generated by depolarizing membrane, but the electrode sees the moving front of polarization, rather than the direct charge flow between regions of depolarization and repolarization. Most EPs are far-field recordings, because the generators are so deep that electrodes can not obtain a near-field recording. Therefore, a potential complex may consist of individual waves generated by widely separated anatomic structures. The best example of this is the AEP; conduction in a complex pathway through the brainstem results in a multiwave complex that is less than 10 ms in duration.

The amplitude of far-field potentials is small, because of the distance between the generator and the recording electrodes. However, the potential is less susceptible to attenuation by scalp and other tissues. The potential field is distributed widely.

### 1.3.4.3 Mixed near-field and far-field recordings

AEPs and SEPs have waves generated by far distant structures, so an individual complex can consist of a combination of near-field and far-field potentials. This is especially true with recordings that employ distant noncephalic reference electrodes.

### 1.3.5 Stimulus Rate

#### 1.3.5.1 Transient EPs

Transient EPs are generated by stimuli so widely spaced in time that the response to one trial is completed before the beginning of the next. Long-latency EPs, such as the VEP, can be repeated at rates of no more than 1–2/sec. Short-latency EPs can be presented at rates up to 10/s. Transient EPs are the most commonly used clinical EPs.

#### 1.3.5.2 Steady-state EPs

Steady-state EPs are created by stimuli presented at a rate fast enough that the EP becomes a rhythmical wave with the same frequency as the stimulus. Responses at harmonic and subharmonic frequencies may also be seen. Steady-state VEP is obtained with a frequency of about 5/s. The short-latency steady-state AEP is obtained with a frequency of about 250/s.

### 1.3.6 Stimulus Duration

Stimulus duration differs markedly among modalities. The duration of the electric shock for an SEP is often 0.2 msec (200 µsec). In contrast, the VEP is usually generated by a pattern-reversal stimulus that has a period of 500 msec. When the duration is long, an EP can be a composite of the responses to the onset and offset of the stimulus. This is particularly true for the pattern-reversal VEP and click AEP. The responses to each type of stimulus are averaged separately. The short duration of the SEP stimulus makes onset and offset effects negligible.

### 1.3.7 Unilateral and Bilateral Stimuli

AEPs and SEPs test one side of the body at a time. VEPs usually test one side, but partial field stimulation is occasionally performed; this involves stimulation of the homonymous left or right or upper or lower fields from both eyes.

Bilateral stimulation produces a response that differs from the sum of the independent unilateral EPs. Not only are there the additive effects of the potentials, there also is interaction between the sides, so that activation of one side may influence the response to stimulation of the other side.

### 1.3.8 Midline versus Lateral Recording

VEPs are usually obtained by midline occipital recording electrodes. Lateral electrodes are sometimes used, especially when there is concern about a retrochiasmatic lesion. AEPs use vertex and ear electrodes, so the recording is a montage of midline and lateral electrodes.

## 1.4 THE GENERATORS OF EPS

Localization of the generators of EPs is difficult. It is attractive to think of a complex waveform as being composed of individual waves, each generated by a specific neural structure along the pathway through the nervous system. Unfortunately, this is rarely the case. Charge movement through the three-dimensional brain can produce a complex waveform even with a single generator. Multiple generators further confound this effect.

In general, if the response is of high amplitude and localized distribution, the generator is near the recording electrode. If the response is low amplitude and of wide distribution, the generator is likely to be distant and usually deep in subcortical tissue.

Localization is based on wave amplitude and conformation, as well as neuroanatomy, but the most important information is frequently obtained from lesion studies. Studies involve either animals with experimental lesions, or patients with naturally occurring lesions. The anatomic defect is known from autopsy or imaging.

So far, the generators of most EP peaks have been determined only approximately. There is much debate about the brainstem generators of each of the peaks of the AEP. For the SEP, there is disagreement about which peak signifies arrival of the impulses at the cortex. Similarly, the relative contribution of the primary and secondary visual cortex regions to the VEP is controversial.

Numerous studies have investigated the generators of EPs. Three types of generators appear to contribute to the recordings: (1) cortical EPS, (2) subcortical EPs, and (3) EPs recorded from peripheral nerves.

- Cortical EPs are largely due to the spatial and temporal summation of excitatory and inhibitory postsynaptic potentials generated at the membranes of nerve cell bodies and dendrites in response to the input produced by the stimulus. The potential fluxes create currents that penetrate the cortical surface, skull, and scalp, thereby producing electric fields detectable by surface electrodes (Figure 1.1).

- Subcortical EPs are probably a mixture of two components: (1) postsynaptic potentials generated in groups of neurons of subcortical relay nuclei, and (2) action potentials of the connecting axonal tracts. The first component, consisting of stationary generators of electric fields, is probably responsible for those subcortical potentials that can be recorded with similar latency at various distant electrode sites. The second component, consisting of propagated waves of depolarization, may explain why some subcortical EPs appear with delays of up to a few milliseconds at different recording sites.
- EPs recorded from sensory nerves are due to a wave of depolarization propagated along the membrane of the nerve fibers. When passing under a stationary recording electrode on the skin, the wave produces a major surface-negative deflection that may be preceded and followed by minor positive deflections due to the approaching and disappearing wave. The compound action potential may include later deflections generated by fiber groups of lower conduction velocity, but these deflections are of low amplitude, due to the greater temporal dispersion of slowly conducted impulses.

For all three kinds of EPs, the shape, size, and timing of an EP recorded from the scalp or skin depends on many factors, including the duration of the potential change, and the size and spatial orientation of the generator to the recording electrodes.

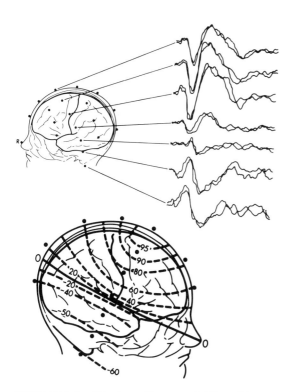

FIGURE 1.1. *Plotting of isopotential lines. Top: AEPs recorded from each of the designated electrodes referenced to the nose. Bottom: Isopotential lines indicating zones of similar amplitude of the major peak.*

# 2

# General Methods of Stimulating and Recording

## 2.1 STIMULUS METHODS

### 2.1.1 Stimulus Types

EPs differ not only depending on stimulus modality, but also depending on type of stimulus. For example, the visual evoked potential (VEP) produced by checkerboard pattern reversal differs from that produced by diffuse light flashes. There are even differences in EPs that depend on the specific method of generating the stimulus. For example, the VEP to checkerboard pattern reversal may differ depending on whether the pattern is generated by a TV screen, an array of light-emitting diodes (LEDs), or projection using a moving mirror. AEPs to clicks differ from those to tones, and AEPs to clicks differ depending on whether the clicks move the air in the external auditory canal toward or away from the eardrum. Only SEPs, generated by direct electrical stimuli, are less capricious.

### 2.1.2 Stimulus Intensity

In general, a stimulus must have a minimum intensity to produce an EP; this is the *threshold*.

This minimum may be difficult to determine because EPs to weak stimuli have low amplitude and may require averaging of a great number of responses. An increase of stimulus intensity above threshold usually increases the amplitude and may decrease the latency of the EP; it may also change the shape and add more peaks to the EP. Amplitude does not usually grow linearly with stimulus intensity but may increase in proportion to the power or logarithm of stimulus intensity. Amplitude increases with stimulus intensity to a maximum, after which it is usually stable but may decrease slightly if the stimulus is increased further. The latency of an EP may reach a minimum at stimulus intensities different from those that give maximum amplitude of the peak. Latency and amplitude of each peak of an EP may vary independently of other peaks.

### 2.1.3 Stimulus Duration

An increase in the duration of short stimuli has an effect similar to that of an increase of stimulus intensity. Such an increase therefore often increases the amplitude and decreases the latency of EP peaks. A long-duration stimulus may produce separate EPs at stimulus onset and offset.

### 2.1.4 Stimulus Rate

Stimulus rate accounts for the qualitative difference between transient and steady-state EPs. Specific stimulus frequencies producing transient and steady-state EPs depend on the EP type.

Stimulus rate is measured in cycles per second or hertz (Hz) for sine wave stimuli such as sound waves, and in number per second for other stimuli. The denomination *Hz* is sometimes used to indicate stimulus rate (10 Hz vs. 10/sec) but this is improper use; *Hz* should refer only to sinusoidal stimuli.

### 2.1.5 Number of Stimuli

The number of stimuli to be used for an average depends on the amplitude of the EP components relative to the amplitude of the ongoing EEG activity.

### 2.1.6 Size and Location of Stimulated Area

VEPs differ depending on the stimulated part of the visual field. AEPs vary with the frequency of the sound stimulus that excites different parts of the cochlea. SEPs to stimulation of different arm and leg nerves have different latencies, shapes, and distributions.

## 2.2 RECORDING METHODS

### 2.2.1 Electrode Types and Applications Methods

#### 2.2.1.1 EEG electrodes

EEG electrodes are used most commonly for EP recordings from the scalp and often also for EP recordings from the skin. They consist of a metal cup, often having a central hole and a flat rim with an outer diameter of 4–10 mm. The cup is attached by an insulated lead wire to a connector plug. The site of electrode application is determined by measurements referenced to bony landmarks. The skin is prepared by wiping

with alcohol or acetone; a mild abrasive may be used to remove dry superficial skin layers.

Several methods may be used to attach electrodes to the scalp and skin. The collodion technique gives stable recordings with few artifacts. In this technique, a cup electrode with a central hole is placed on the prepared application site and held in place with a stylus while collodion is applied around the edge of the electrode and spread onto the scalp or skin. Spreading and drying of the collodion are facilitated by a stream of compressed air guided through a tube around the stylus. After the electrode is securely attached, the stylus is removed and the cup is filled with conductive jelly injected through the central hole. Electrodes are removed by dissolving the collodion with acetone. Because of the chemicals used, this method should not be used in areas with limited ventilation or with explosion hazard, such as infant isolettes or operating rooms.

Another method of applying electrodes uses a conductive paste that can both hold the electrode in place and provide good electric contact. A dab of paste is placed on the prepared site and a cup pressed down until it makes contact with the skin. The electrode is covered with a gauze pad to delay drying of the paste and facilitate attachment. After the recording, the electrodes can be easily removed and the paste washed off with water. Although this method is faster, it is susceptible to more mechanical and electric problems than the collodion technique.

#### 2.2.1.2 Clip electrodes

An electrode mounted in a clip is sometimes used for recordings from the ear lobe. This electrode should consist of the same kind of metal as that of scalp electrodes so that it can be paired with those electrodes at the inputs of the same amplifier without degrading the signal. Many patients find these electrodes uncomfortable, so conventional electrodes placed behind the ear should be used for most applications.

#### 2.2.1.3 Adhesive EKG electrodes

For recording the EKG, metal electrodes covered with conductive gel may be attached to the skin with an adhesive patch and should be paired with similar composition electrodes at the amplifier input.

### 2.2.1.4 Needle electrodes

Needle electrodes are used for recording in some laboratories. Sharp steel or platinum wires are inserted into the superficial layers of the skin or scalp after disinfection of the insertion site. The lead wires of the needle electrode must be attached to the skin or scalp to avoid pulling the needles out. Although they can be applied rapidly and usually cause no problems, these electrodes carry the risk of discomfort, infection, and electric recording problems and should be used only under restricted conditions. Because they have high impedance, especially for low frequencies, they may be useful for recordings of fast EPs with amplifiers of high input impedance. Needle electrodes are occasionally used for electrocochleographic recordings, for SEP recordings from the interspinal ligament, and for recordings of sensory nerve action potentials from peripheral nerves. Because of the risk of infection, needles should not be used during prolonged recordings, such as those needed for monitoring in intensive care units. They should be discarded after use in patients who have, or may have, Jakob-Creutzfeldt disease.

### 2.2.1.5 Electrocochleographic electrodes

For transtympanic recordings, a needle electrode may be placed through the eardrum, so that the electrode tip lies against the promontory of the middle ear near the cochlea. These electrodes should be inserted only by otological specialists using general anesthesia in children and local anesthesia in adults. In extratympanic recordings, the electrocochleogram is recorded from a needle electrode inserted into the anterior wall of the external auditory canal or from specially shaped electrodes placed on the lumen of the ear canal.

### 2.2.1.6 Electroretinographic electrodes

Recordings of the electroretinogram (ERG) can be made from the eyeball with a contact lens attached to a scleral speculum or with a light-weight corneal contact lens. The lenses contain one or two recording leads and usually degrade visual acuity to some degree but may permit stimulation with patterned light. For unobstructed vision during the recording, a gold foil electrode may be hooked over the lower eyelid or minute silver-impregnated nylon fibers may be placed on the cornea. Periorbital EEG electrodes are less suitable for ERG recording.

### 2.2.1.7 Electrocorticographic electrodes

Recordings from the cortical surface may be made during neurosurgical procedures exposing the cerebral cortex. The electrodes consist of spring-mounted metal balls or saline-soaked cotton wicks. A matrix of metal electrodes embedded in a sheet of plastic material may be used for chronic subdural recordings.

### 2.2.1.8 Intracerebral electrodes

Multicontact wire electrodes may be inserted stereotactically into the brain for acute recordings during an operation or for subsequent chronic recordings.

## 2.2.2 Electric Properties of Recording Electrodes

Recording electrodes should be capable of conducting, without distortion, the potential changes in the frequency range of the EP to be recorded. To obtain this recording capability, one must

- choose recording electrodes of suitable material,
- test electric continuity, if in doubt, by measuring the electrode resistance between the ends of the electrode,
- evaluate the electric contact between electrodes and the scalp or skin by measuring electrode impedance, and
- avoid electrode polarization and bias potentials.

### 2.2.2.1 Electrode materials

The materials on the electrode surface should not interact with the electrolytes of the scalp or skin. Electrodes coated with gold, tin, or platinum are satisfactory. Silver electrodes coated with silver chloride are required for recording potentials that are slower than those recorded in routine clinical work.

### 2.2.2.2 Electrode resistance

Resistance, or opposition to *direct current* flow, is measured to test the electric continuity of an electrode if a break in continuity is suspected. For this measurement, the two ends of the electrode are connected to an ohmmeter which passes a weak direct current through the electrode and shows a readout of resistance; the resistance of an intact electrode measures no more than a few ohms.

### 2.2.2.3 Electrode impedance

Electrode impedance, or opposition to *alternating current* flow, is measured to ascertain good electric contact between an electrode and scalp or skin after the electrode has been applied. This measurement should be made before the start of every recording and should be repeated during the recording if there is reason to suspect bad electrode contact. The impedance of an EEG scalp electrode should be between 1,000 and 5,000 ohms.

Electrode impedance is measured with an impedance meter, which passes a weak alternating current from the electrode selected for testing to all the electrodes connected to the meter. Alternating current is used for this measurement because it is more representative of the alternating potentials in the EP than direct current and because it avoids electrode polarization caused by direct current.

Both very low and very high impedance are undesirable. Very low impedance short-circuits the amplifier input and is often due to smear of electrode gel or sweat; electrodes with impedances less than 1,000 ohms should be inspected, cleaned, and reapplied. An electrode of very high impedance paired with an electrode of lower impedance may cause an imbalance at the inputs of a differential amplifier which favors the recording of interference, especially 60 Hz artifact from power lines. Only when the electrode impedance is so high that it equals or exceeds the input impedance to the amplifier will it significantly reduce the amplitude of the recording. Electrodes with impedances over 5,000–10,000 ohms should be checked and usually need improvement of their mechanical and electric contact with the scalp or skin, although a break in continuity, such as a break between electrode cup and lead wire or between lead wire and plug, may also cause very high electrode impedance.

### 2.2.2.4 Electrode polarization and bias potentials

Electrode polarization and bias potentials may distort EP recordings but are easily avoided with modern techniques. Polarization is caused by the flow of electric current through the recording electrode. The current distributes ions at the electrode so that current flows better in one direction than in the other and thereby distorts the recording of alternating potentials. Polarization can be minimized by measures that reduce the flow of current through recording electrodes, namely, by using amplifiers with high input impedance and electrodes with fairly large contact areas, and by avoiding steady current flow, especially that used to measure electrode resistance while the electrode is in contact with scalp or skin.

Bias potentials are caused by ion exchanges not due to current flow. These potentials can be avoided by using pure metal electrodes with clean surfaces and by pairing similar electrodes at the inputs of each amplifier.

## 2.2.3 Electrode Placement

The placement of recording electrodes depends on the EP type to be recorded and generally aims at obtaining the highest amplitude and clearest definition of the peaks of an EP. Electrodes are placed as closely as possible to the presumed generator in near-field recordings; widely spaced electrodes are used in far-field recordings from distant structures.

To make electrode placements constant, the site of each electrode must be determined by measuring coordinates with reference to standard landmarks such as bridge of the nose, mastoid process, parts of the ear, cervical or lumbar vertebrae, and clavicle. A tape measure and marker pen should be used in each case. The International 10–20 System, widely used in clinical EEG recordings, may be employed to determine the positions of the vertex and midfrontal electrode locations often used in EP recordings. The 10–20 system may also be used to designate the position of other scalp electrodes by indicating their distances relative to the points defined by the 10–20 system.

## 2.3 AMPLIFIERS

In many modern averagers, the amplifiers, input selector switches, calibration units, and filters are housed with the computer in a single unit; only the input board is separated from the other electronic apparatus so that it can be brought near the patient. The patient electrodes are connected to the input board.

### 2.3.1 Input Board

The input board is a box that contains receptacles for the electrode connector plugs. Output of the board is delivered to the amplifier by a shielded cable. The box usually contains circuitry that amplifies the signal to reduce the effect of noise encountered during conduction to the main amplifier. This circuitry also should limit current flow to 20 microamps per electrode, thereby protecting the patient from electric shock. The box also usually allows for testing of electrode impedance.

The receptacles are usually labeled either numerically in pairs (e.g., Input 1, Input 2), or according to position (e.g., Cz, A1). They are connected to selector switches that are used to select two electrodes as inputs to each amplifier.

The input board also provides a receptacle for the ground electrode. A ground electrode on the patient should always be connected to this receptacle unless the subject is grounded through another connection, for instance, through a ground connection to a stimulator or EKG monitor, in which case the ground of the recording equipment must be connected to the

ground of the other equipment. Grounding is most effective in reducing interference if the ground connection is close to the recording site, therefore, it may be necessary to use a ground electrode placed near the recording electrodes for EP recordings and to disconnect the patient from the ground lead of the other equipment.

### 2.3.2 Calibration

The recording system must be calibrated at the start of each recording session. This is done by feeding square-wave pulses of known voltage to the inputs. These pulses must be amplified with the same gain and filter settings, averaged for the same number of trials, and displayed with the same gain settings as the responses recorded from the nervous system. This calibration procedure requires a source of selectable, very precise electric pulses that can be driven to appear at a constant interval after a trigger pulse. The selection should include pulses of 5 and 10 $\mu$V and of 10–50 msec for cortical EPs and of 0.5 and 1.0 $\mu$V and of 2–5 msec for subcortical EPs. Each channel is tested. The horizontal deflection of the averager display usually does not need to be calibrated because the timing circuits of modern computers are extremely precise.

Virtually all EP machines have cursors that can be placed on any point of the computer display to give a readout of amplitude and latency. Latency and amplitude differences between cursors are then calculated by the computer. The averaged calibration pulse may be used to verify the amplitude readout of the cursors: When one cursor is placed on the calibration pulse and the other on the baseline, the readout should equal the voltage of the averaged pulse. In older averagers not equipped with cursors, a hard copy of the calibration pulse is made with the same gain settings as used for copies of EPs; the height of the calibration pulse is measured and compared with the height of EP deflections. For instance, if a calibration pulse of 10 $\mu$V has a height of 5 cm, and an EP peak has an amplitude of 2.5 cm when displayed in the same manner, then the peak has an amplitude of 5 $\mu$V.

### 2.3.3 Input Impedance

The electric impedance of the amplifier input must be high compared with that of the recording electrodes to avoid loss of amplitude of the signals recorded. The input impedance of the amplifiers should be 10 megohms or more.

### 2.3.4 Differential Amplification and Polarity Convention

Amplifiers have several functions. The most obvious is to increase the amplitude of the biological signal. The *differential amplifier* compares the two inputs and amplifies the difference between them. Therefore, the differential amplifier rejects cerebral and extracerebral potentials common to both inputs.

#### 2.3.4.1 Discrimination of cerebral potentials and polarity convention

One electrode is connected to each of the inputs of the differential amplifier, called *input 1* and *input 2*. The differential amplifier subtracts the voltage of input 2 from that of input 1, then amplifies this difference. This results in amplification of mainly cerebral potentials, which usually have different voltage and timing at the two electrodes. Extracerebral potentials, including 60-Hz interference and other artifacts, usually have similar timing and amplitude at the two electrodes, and are therefore cancelled at the amplifier input.

Unfortunately, interpretation of the output of a differential amplifier is more complicated than that of a single-ended amplifier. A positive output wave means that input 1 is positive with respect to input 2. This can be either from positivity at input 1 or negativity at input 2; there is no way to differentiate between these possibilities on a single channel recording. In many recording situations, a particular wave may be a combination of potential changes at both electrodes.

Display of the potential depends on waveform convention. For EEG, convention is that negativity at input 1 produces an upward deflection (Figure 2.1). EPs do not share this consensus; the trace is usually displayed so that the

wave of interest is convex upward. American EEG Society Guidelines for polarity convention are presented in the sections on the respective EPs. Because of the potential for confusion, the polarity should be indicated on the scale of the hard copy.

The results of differential amplification using either polarity convention are illustrated in Figures 2.2 and 2.3. Figure 2.2 shows the effect of signals of different amplitude and polarity; Figure 2.3 shows the effect of signals of different timing and polarity. Both figures acknowledge the fact that neither electrode is likely to be entirely unaffected by potentials near the other electrode. In summary, the output of the amplifier always indicates the net difference between the inputs; the polarity of the output depends on the polarity convention of the specific amplifier; and the amplitude depends on the gain setting.

The cerebral potential changes reflected in EPs rarely come from only one electrode, although this could be the case in a recording using an electrode located on the scalp directly over a small area of cortex generating the EP and another electrode located at a great distance from the generator. In these rare instances, the first electrode may deserve the

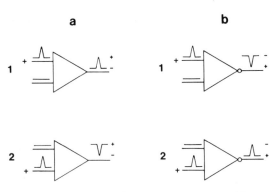

FIGURE 2.2. *Two polarity conventions used in EPs. Convention a is the same as for EEG, where a positive potential at input 1 produces an upward deflection at the output. Convention b is opposite to a, positivity at input 1 produces a downward deflection. For both conventions, a negative input at input 2 produces the same deflection as a positive deflection at input 1.*

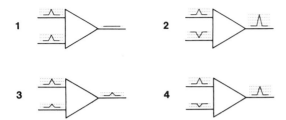

FIGURE 2.3. *Properties of a differential amplifier. For each example, the signal delivered to input 2 is subtracted from the signal delivered to input 1. Amplification is ignored. (1) Signals of identical polarity and amplitude. (2) Signals of opposite polarity but identical amplitude. (3) Signals with the same polarity but different amplitudes. (4) Signals with different polarity and amplitudes.*

name *active* or *exploring electrode* and the other *inactive, indifferent,* or *reference electrode*. However, the terms *active* and *reference* are often used less strictly to indicate that one electrode is closer to the generator than the other one. The term *referential recording* is sometimes used to describe this recording situation. The terms *monopolar* and *unipolar recording,* occasionally used to suggest that signals originate only at one electrode, should not be used in any description of differential recordings because all such recordings are made between two points.

In most EP recordings, both electrodes pick up some potential changes so that each electrode adds or subtracts parts of the EP; the contribution by each electrode depends mainly on its distance and spatial orientation with respect to the generator. The distinction between active and reference electrode becomes unimportant. It loses its meaning entirely in far-field recordings in which neither electrode is located much closer to the generator than the other one. The term *bipolar recording* is sometimes used to denote recordings between two electrodes that are both known to contribute to the EP. However, the terms *bipolar* and *referential* are best reserved for multichannel recordings.

The term *average reference* electrode denotes recordings attempting to find an average potential level against which to record potential changes at another electrode. This average is produced by using more than one electrode as input 2 of an amplifier so that each electrode selected contributes equally to the average potential at that input. For instance, electrodes on both ears or both mastoids may be connected to one input and used as an average.

Simultaneous recordings of EPs in more than one channel are made by selecting different electrode pairs as inputs to each channel, creating different montages. Two types of montages are used in multichannel recordings:

- Referential montages use the same electrode as input 2 of all channels; this electrode, usually located at some distance from the areas generating the EP, becomes the reference for the various electrodes connected to input 1 of all channels.
- Bipolar montages connect different pairs of electrodes to the different amplifier inputs; often the electrode at input 2 of one channel is also used at input 1 of the next channel so that the electrode pairs of successive channels form a chain along or across the head. Both electrodes in a pair contribute to the EP, each to a different degree.

### 2.3.4.2 Rejection of artifacts

It should be understood that differential amplification rejects only those artifacts that

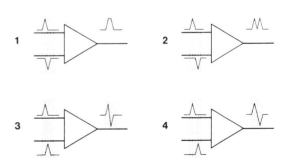

FIGURE 2.4. *Effect of timing of the input signal on output of a differential amplifier. This figure shows that a complex waveform can be the result of addition of elemental waveforms. (1) Prolonged waveform. (2) Bifid peak. (3) Biphasic waveform. (4) Biphasic peak with prolonged duration.*

cause identical potential changes at both inputs; such potential changes are said to be *in common mode*. Other artifacts, greater at one electrode than at the other, are amplified like cerebral activity. By the same rule of differential amplification, cerebral activity, such as the EEG or the responses averaged in EPs, is rejected to the extent that it has the same polarity, amplitude, and timing at both inputs. The ability to reject common mode signals is expressed as the *common mode rejection ratio*. This is the ratio of the amplifier output produced by a signal applied differentially, that is, between the two inputs, over the amplifier output produced by the same signal when it is applied in common mode, that is, between both inputs tied together and the amplifier ground. The common mode rejection ratio of amplifiers used for EP recordings should be 10,000:1, that is, 80 dB or more.

Effective rejection of artifacts depends on several factors. A defect in any one of these factors is often revealed by the appearance of 50–60 Hz alternating current (AC) artifact in the recording. This artifact is due to interference from power lines. This interference is present to some degree at all EP recording sites; in some areas, interference may be strong enough to require electric shielding of the recording room. The artifact has a frequency of 60 Hz in the United States and of 50 Hz in countries using alternating current of that frequency. Line interference is usually removed by the differential amplifier because it usually appears with similar amplitude at the two inputs. However, various defects of the recording electrodes may make the interference appear with different amplitude at the two inputs. This amplitude difference is amplified and appears at the output.

The most common cause of 60-Hz artifact is unequal electrode impedance. This is often due to partial or complete loss of contact between one electrode and scalp or skin or to a break in the continuity of an electrode, for instance, a break between lead wire and metal cup or connector plug. Infinitely high impedance may result from the inadvertent failure to connect an electrode to an open input and has a similar effect as grounding one input: The amplifier then no longer operates differentially but increases signals between the remaining input and the ground electrode. Problems with the ground electrode are a frequent cause for picking up 60-Hz interference in differential recordings. If the ground is not connected to the subject, the amplifier inputs float without reference to the potential level at the recording site, and the amplifier uses its own internal ground potential level to subtract signals at input 2 from those at input 1. Input potentials that are equal with reference to the subject's head, and would therefore be eliminated by differential recording with a proper ground connection to the head, may now appear different with respect to the amplifier ground and not be cancelled. For instance, 60-Hz or heart-beat artifacts may not be nulled. Cerebral potentials may show little obvious distortion when recorded with floating inputs. However, electric hazards for the subject are generally increased in the absence of a ground connection between subject and recording equipment.

If only one amplifier input is connected to the subject, that is, if both the other input and the ground connection are open, the output of the amplifier represents the potential difference between the electrode on the subject and the amplifier ground; this kind of recording, although at times resembling a normal recording for brief periods, is meaningless and usually marred by artifacts.

The electronic parts of the amplifier themselves give rise to potential fluctuations that are added to the output generated by the input signals. The noise level of the amplifier, measured when both inputs are closed and the filters wide open, should not exceed a few microvolts.

### 2.3.4.3 Amplification

*2.3.4.3.1 Gain and sensitivity.* The increase of the voltage of a signal between the input and the output of an amplifier can be described by *gain* or *sensitivity*. Gain is the ratio of signal voltage obtained at the amplifier output to the signal voltage applied at the input: An amplifier set to give an output signal of 1 V for an input signal of 10 µV has a gain of 100,000. This voltage ratio is sometimes expressed in decibels, the number of decibels amounting to 20 times the logarithm$_{10}$ of the voltage gain (dB = 20 × log$_{10}$gain). For instance, a gain of 10 equals that of 20 dB. In contrast, sensitivity is the ratio of input voltage over the size of the deflection it produces in a tracing of the output. For instance, an amplifier set to produce a vertical deflection of 1 cm for an input of 10 µV has a sensitivity of 10 µV/cm, or 1 µV/mm.

*2.3.4.3.2 Recording and display gains* The differential amplifier at the input of an averager increases the voltage of the signal as it is being recorded. Another amplifier is used to display the output of the computer, namely, the digitized EP, after the completion of averaging. Although both the recording and the display gain affect the same substrate, the vertical size of the tracing, they do so at different stages of the recording and may substitute for each other only within limits. In particular, the recording gain must be set to match the vertical resolution of the analog-to-digital converter for the most effective averaging. The display gain is used to adjust the size of the completed average for viewing and plotting purposes.

## 2.4 FILTERS

Filters serve to exclude from the recording those potential changes that have frequencies different from the frequencies represented in the response under study. This form of signal conditioning reduces unwanted waveforms before they reach the computer. The noise-reducing, signal-enhancing action of filters resembles the effect of averaging and indeed can reduce the number of responses that need to be collected for the clear definition of an EP. Even though averaging alone can enhance the EP, without filtering, the number of responses required for clear definition of the EP may be impractically large. Also, filtering of high-frequency components is needed in conjunction with averaging to eliminate components that exceed the limit imposed by sampling rate.

Filtering can be accomplished in several ways. The most commonly used filters reduce or eliminate frequencies higher or lower than a middle range that contains the frequencies of the EP components under study. Most filters operate on the recording before it is digitized and are said to be *analog filters;* they can be divided further into *passive filters* made of resistive and capacitative components and *active filters* containing power-consuming components such as transistors. Analog filters not only reduce the amplitude of unwanted components, but also distort the timing relationship between some of the desired signal components passing through the filter, especially those near the cutoff frequencies. This phase shift can be minimized, and a very sharp cutoff can be accomplished by the use of some types of *digital filters.*

The effect of high- and low-frequency filters is usually specified in terms of the frequency of sine waves that are reduced in amplitude by a certain fraction. The frequency range between the low and high cutoff points, which is not significantly affected by the filter settings, is called the *bandwidth,* or *bandpass.* Because most EPs do not have the shape of sine waves, the choice of the most suitable filter settings may be difficult. A very rough estimate of filter action can be made by measuring the lengths of the waves in an EP. For instance, BAEPs contain wavelets lasting about 1–2 msec, corresponding with frequencies of 500–1,000 Hz. However, filter settings excluding all but this narrow range would not be satisfactory because BAEPs also contain slower and faster rising and falling phases that would be deformed by filtering unless wider frequency settings were used. A good practical estimate of filter action can be obtained by comparing the effects of different filter settings on the same EP; for this purpose, an EP may be recorded simultaneously in different channels with different filter settings, or the response may be stored and played back repeatedly for averaging with different filter settings. Although such an estimate can not be made for each recording, estimates for each type of EP have been obtained often enough to provide practical guidelines for suitable filter settings for each EP type. Theoretically, the effect of filters can be predicted by computing the power spectrum of the EP, which shows the spectrum of frequencies that are contained in the EP and thereby indicates the frequencies that may be eliminated by filtering.

## 2.4.1 Analog Filters

### 2.4.1.1 Low-frequency filter

Low-frequency filters (LFF) reduce the amplitude of slow waves without attenuating faster waves. The LFF has also been termed *high-pass filter,* however, this term is not preferred. The action of the LFF can be described in terms of the low filter frequency setting or the time constant. The low filter frequency indicates the frequency of sine waves that are reduced in frequency by a specified fraction. The effects of different low-frequency filter settings on sine waves of different frequencies are shown in Figure 2.4 with dotted lines. In the commonly used passive filter made of a capacitor (C) in series and a resistor (R) in parallel to the signal path, the cutoff is usually defined as the 3 dB point,

that is, the frequency at which a signal at the filter output is reduced by 3 dB, or 30%; this frequency (f) amounts to:

$$f = \frac{1}{2\pi RC}$$

Slightly higher frequencies are affected to a minor degree; lower frequencies are progressively more reduced. This progression has a roll-off slope of 6 dB/octave, that is, the amplitude drops by 50% as the frequency is reduced by a factor of 2. Passing the signal through another filter of the same setting, or using the same filter twice, for instance, during tape recording and playback, doubles the steepness of the filter slope. Very steep slopes can be obtained with active filters. For clinical EP recordings, roll-off slopes for LFFs should not exceed 12 dB/octave because steeper slopes may cause too much phase shift.

The time constant (T) is the time required by a sudden sustained voltage change to decay to

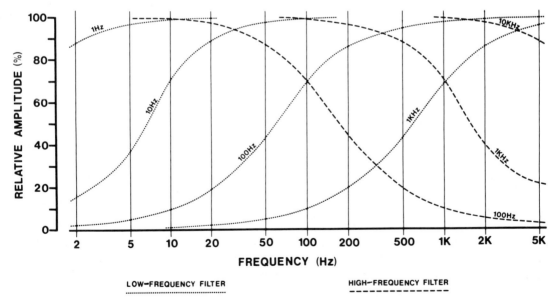

FIGURE 2.5. *Frequency response of low- and high-frequency filters. Horizontal axis is signal frequency. Vertical axis is relative amplitude of the amplitude output. Response of low-frequency filters is indicated by dotted lines, high-frequency filters by dashed lines. The turnover frequency for each filter setting is indicated on each line. At the turnover frequency, the amplitude is reduced by about 30%. The combination of low- and high-frequency filters defines the bandwidth.*

37% of its peak value. The time constant of a simple resistor-capacitor (RC) filter is:

$$T = R \times C$$

where R is the resistance of the resistor in ohms and C is the capacitance in farads. The time constant is related to the cutoff frequency (f) by the equation:

$$T = \frac{1}{2\pi f}$$

The time constant may be determined by measuring the time interval between the onset of a sustained calibration pulse and the point at which its amplitude has dropped to 37%. The 37% figure is derived from the reciprocal of *e* or approximately 1/2.7.

The limits of low frequency vary with the design of the coupling of amplifiers. Coupling through low-frequency filters limits the recording to alternating signals such as AC. To record sustained potential changes such as generated by direct current (DC) requires direct coupling. Amplifiers used in routine clinical EP recordings should have LFF settings that include 0.1 Hz. Lower filters or direct coupling (DC recording) may be needed for very slow poten-

tial changes such as those of some of the event-related potentials.

### 2.4.1.2 High-frequency filter

High-frequency filters (HFF) reduce the amplitude of waves of high frequency and let waves of low frequency pass without attenuation. An older name is *low-pass filter*; use of this term is discouraged. The action of different high-frequency filter settings is shown in Figure 2.4 with dashed lines. This action is usually specified by the HFF which is the mirror image of the LFF, having 3-dB cutoff points at

$$f = \frac{1}{2\pi RC}$$

and roll-off slopes of 6 dB/octave for filters made of a resistor in series and a capacitor in parallel with the signal path. The time constant, or the time required for a sudden input voltage to rise to 63% of its peak value, equals $T = R \cdot C$ in an RC filter.

Amplifiers to be used for clinical EP recordings should have HFF settings including at least 5,000 Hz. HFFs should have roll-off slopes not exceeding 24 dB/octave.

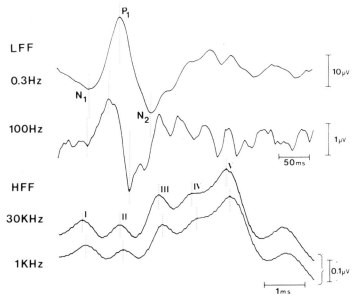

FIGURE 2.6. *Phase shift produced by filters. Top two tracings: Effect of two different low-frequency filter settings on a VEP recording. Increasing the low-frequency filter setting results in a shift of the major positive peak to the left, that is, the appearance of a shorter latency. Bottom two tracings: Effect of two different high-frequency filter settings on a BAEP recording. A decrease in the high-frequency filter setting results in a shift of the waves to the right, that is, the appearance of a longer latency.*

### 2.4.1.3 60-Hz (notch) filter

Some amplifiers provide a filter that greatly reduces the amplitude of waves or frequencies in a very narrow band centering at 60 Hz, the most common artifact from power-line interference in countries using AC of 60 Hz; a 50-Hz notch filter is used in countries with that power-line frequency. Although this filter affects only a very narrow part of the spectrum of frequencies in an EP, the frequencies of the filtered band may form an important part of the EP, and the EP may be distorted significantly by the use of the filter. The filter should therefore not be used routinely. If it must be used, the same responses should be averaged without the filter to assess how useful the filter was in reducing the artifact and how harmful it was in distorting parts of the EP.

### 2.4.1.4 Phase shift

Analog filters alter not only the amplitude but also the timing of signals. LFFs affect mainly slow waveforms and make them appear earlier than fast ones (phase lead); HFF are of importance for fast deflections and make them appear later than slow ones (phase lag) (Figure 2.5). The phase lead and lag induced by a passive RC filter at different low and high cutoff frequencies, respectively, are shown in Figure 2.6.

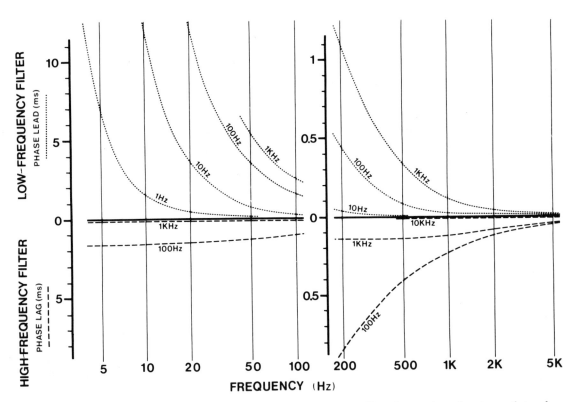

FIGURE 2.6.  *Effect of low-frequency filter (top) and high-frequency filter (bottom) on the phase of signals of different frequency (horizontal axis). Values above the horizontal zero line indicate the amount of time by which the low-frequency filter shifts signals of low frequency ahead of faster frequencies. Values below the zero line indicate the amount of time by which high-frequency filters shift signals of high frequency behind those of lower frequency. These curves were derived from measurements of the filters of the same amplifier used in Figure 2.4. Comparison of the two figures shows that when a particular frequency is affected by phase shift, the amplitude is reduced as well.*

Although these effects are most prominent at frequencies near the cutoff and beyond, they extend into the range of the frequencies that are passed with little attenuation and may significantly distort the timing of EP peaks. In particular, phase shifts of either kind may falsify the interval between the stimulus and the EP peaks and may distort the time interval between EP peaks, especially between peaks composed of different frequencies. Passive filters generally shift phase less than active filters, although the degree of phase shift varies with the type of active filter.

## 2.4.2 Digital Filters

Digital filtering consists of the use of computer programs that operate either on the digitized responses before averaging or on the EP after completion of averaging. A relatively simple digital filtering method consists of smoothing. Other methods are more complex and have rarely been incorporated in routine clinical averaging procedures. They include methods for elimination of low and high frequencies, Wiener filtering, and time-varying filters.

Computer programs can eliminate components of low or high frequency with very sharp cutoff and may not introduce phase shift. In contrast to analog filtering, some digital filters cause no temporal distortion of EPs. Unfortunately, some of the most commonly used algorithms do introduce phase distortions; choice of filter is crucial.[39]

Wiener filtering separates frequency components of the signal from those of noise either when the characteristics of both components are known beforehand or after they have been calculated from a small sample of EPs. The value of this method is still in doubt.

Time-varying filters are needed if signal or noise is not stationary. Adaptive or learning filters are designed to progressively develop criteria for detection of the signal.

# 3

# Averaging

## 3.1 PRINCIPLES OF AVERAGING

The electric responses of the brain, brainstem, or spinal cord to a single stimulus, when recorded at the surface of the body, are small and obscured by EEG, EKG, muscle activity, and other biological and extraneous electric activity so that individual responses can not be clearly distinguished and seem to fluctuate in repeated recordings. Averaging serves to extract the responses, time-locked to the stimulus and considered signal in this context, from potential changes unrelated to the stimulus and here summarily considered noise. Averaging is done by presenting sensory stimuli repeatedly, collecting and adding each response to the preceding ones. This procedure enhances the signal by reducing the noise toward zero.

Averaging is carried out by a digital computer that

- records electric activity during the selected time period,
- converts the continuous voltage change of analog signal into a sequence of numbers,

- adds the numbers representing recordings of successive responses, then,
- scales them to calculate the average (Figure 3.1).

The period of analysis (epoch or sweep) is started by a trigger pulse from the stimulator and must be long enough to include the response under study. The analysis period is divided into equal periods of time, called *bins* or *points*. The computer samples the voltage then waits until the end of the analysis period before taking another sample. This time is sometimes called the *dwell time*, signifying that the computer dwells on the sample before moving on to the next. This terminology may give the impression that the computer is measuring and perhaps averaging everything happening during that period, but this is not the case. The computer takes its measurement in a minuscule fraction of a second then waits until the designated time to take another sample. A better term is *sampling interval*, which is the time between samples. The sampling interval is the reciprocal of the sampling rate, that is, the number of points sampled each second.

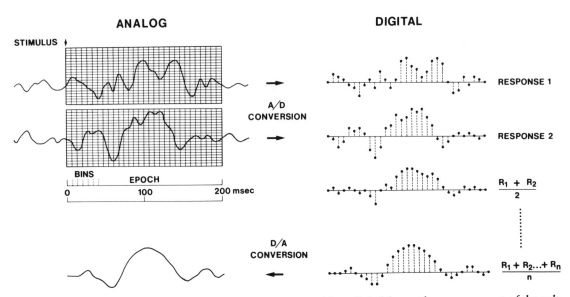

**FIGURE 3.1.** *Averaging of EPs. An analog signal is converted into digital format by measurement of the voltage at defined time intervals, or epochs. The voltage range is divided into discrete voltage increments, or bins, that can be represented by a binary number (top left). Multiple responses are acquired, digitized, and averaged (right). The averaging is performed by adding the discrete voltage values for each of the corresponding epochs of successive responses. The total is divided by the number of trials to arrive at a mean for each epoch. The digital data are displayed on the screen by connecting the dots, making a smooth waveform (bottom left).*

For the analog-to-digital (A/D) conversion, or *digitization,* the amplitude of the recording segment contained in each bin is measured and converted into a number. Because averaging computers operate with digital numbers, the numbers representing amplitude consist of a string of binary digits, or *bits,* either 0 or 1. The result of the A/D conversion of a single response is a sequence of binary numbers that are stored at discrete locations, or addresses, of the computer memory and that can be displayed on the oscilloscope as a sequence of illuminated dots approximating the outline of the response. To obtain an average, successive responses are digitized and added to the preceding ones; after each addition, the numbers in each bin are divided by the number of responses collected to give the running average, until the designated number of responses has been averaged.

All modern EP machines allow simultaneous measurement from multiple channels. One A/D converter can sample multiple channels by systematically sampling each channel in turn during a sampling interval. For example, for a four-channel recording, the computer samples channel 1, then channel 2, then channel 3, then channel 4, then waits for the end of the sampling interval before beginning the cycle again. Obviously, the computer must be able to sample all four channels within the limited time of the sampling interval, but all modern machines have fast enough maximum conversion rates.

## 3.2 COMPUTER TYPES

Two types of computers are used for EP studies, special-purpose and general-purpose. Special-purpose, or hard-wired, computers have one or a few programs permanently stored in their memory. When dedicated to evoked response averaging, these are also called *averaging computers* or *averagers.* Averaging computers usually include amplifiers, filters, calibration signal sources, synchronizing inputs and outputs from and to stimulators, and other options for recording cerebral responses; they may include stimulators. Parameters of stimulation, recording, and averaging can be selected with front-panel controls.

In contrast, *general-purpose computers* are capable of many different tasks and must be programmed for each task by a sequence of instructions given in a specific code, or language. Once a program for averaging has been developed, it can be stored on disk or memory chip and reloaded into main memory when needed. Parameters of analysis are selected by typing on a computer keyboard, using a mouse, or touching screen controls. General-purpose computers are more versatile than hard-wired machines. Cost and setup time are no longer disadvantages with modern machines.

## 3.3 ACQUISITION PARAMETERS

### 3.3.1 Number of Channels

Most averagers have 2, 4, or 8 channels from which the operator can select the number of channels needed for a study. In most averagers, the number of addresses for data is constant, therefore, if more channels are recorded, fewer data points can be recorded for each channel.

The multichannel option may be used to record from different regions of the head or different locations in the body, such as spinal cord or peripheral nerve. Some of the data memory may be used to record previous trials, before it is stored on disk or on tape, but this is temporary and inefficient storage. More memory used for stored data means less available for newly acquired data. Comparing successive EPs from the same region is often done to ensure reproducibility of the recording and to facilitate wave identification.

### 3.3.2 Triggering

To average responses to successive stimuli, each epoch must be started at the same time with respect to the stimulus. In principle, there are three choices for starting points for averaging of transient EPs (Figure 3.2). Not all of them are available on every averager:

- Triggering at the onset of the stimulus (1*a* in Figure 3.2) is the most commonly used method. It includes in the average most of the stimulus artifact and the earliest part of the EP (1*b* in Figure 3.2). In this technique, either the stimulator triggers the computer or vice versa.
- Prestimulus triggering (2*a* in Figure 3.2) includes a period of recording before the

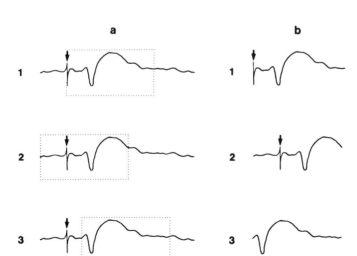

FIGURE 3.2. *Three options for triggering the averager. For each pair,* (a) *shows the recorded potential with the window to be averaged designated by the dotted box, and* (b) *shows the recording that is produced by the averaging using the window to the left of each trace.* (1) *Averaging starts at the stimulus onset.* (2) *Averaging starts before the stimulus to give a prestimulus baseline.* (3) *Averaging starts after the stimulus to avoid having stimulus artifact in the recording.*

stimulus (2*b* in Figure 3.2), which can be used as a baseline against which to measure the amplitude of the EP peaks and as an estimate of the amount of residual noise in the average. Prestimulus averaging may be accomplished either by delaying the stimulus against the synchronizing pulse that starts the computer or by using a computer that retains some data in a buffer before acquisition so that the data can be recalled when a stimulus is triggered.

• Poststimulus triggering (3*a* in Figure 3.2) can be used to eliminate from the EP (3*b* in Figure 3.2) a period between stimulus and onset of the response if this period is very long or contains stimulus artifact. Poststimulus delays are usually generated by the computer, but may be produced by delaying the synchronizing pulse to the computer against the stimulus.

Steady-state EPs are most commonly averaged by triggering the computer at the onset of one of the repetitive stimuli. More than one cycle of the rhythmical steady-state EP is usually included in the epoch. Stimuli occurring during the epoch can not trigger the computer, and only the first stimulus after the epoch is effective in triggering a new sweep. Because the responses to the first few stimuli may undergo adaptive changes before steady-state is reached, collection of steady-state EPs is best started after at least several stimuli have been given.

### 3.3.3 Number of Responses to Be Averaged: Noise Reduction

The number of responses is selected with a hard-wired switch or software selector but the effect is the same. The computer will average the selected number of sweeps. Sweeps rejected because of high-amplitude artifact are not counted. The number of sweeps is determined on the basis of theoretical and practical considerations.

In practice, the number of responses collected should be large enough that successively averaged EPs do not differ from each other. Each EP should be recorded at least twice, and the trac-

ings should be superimposed to ascertain that they resemble each other closely enough in latency, amplitude, and shape. The criteria for replication vary among laboratories but usually amount to a few milliseconds for VEP peaks and to fractions of milliseconds for short-latency AEP and VEP peaks. If two successive EPs do not replicate within these limits, more averages must be obtained until it is clearly established whether an EP of constant features is present. In most cases, satisfactory near-field EPs, such as VEPs, are obtained by averaging 50–200 epochs; far-field EPs, such as the BAEP, require averaging 1,000–4,000 epochs; SEPs are a combination of near-field and far-field recordings and require averaging of 500–2,000 epochs. The number of epochs averaged equals the number of stimuli and of responses in the case of transient EPs. Steady-state EPs, usually averaged for more than one cycle of the rhythmical response, require larger numbers of stimuli.

In theory, the number of epochs required depends on the ratio of the amplitude of the stimulus-locked components in the recording (signal) to the amplitude of the unrelated components (noise). Because averaging reduces the noise, the ratio improves with the number of responses averaged. The relationship between signal-to-noise ratio and number of responses is not linear: The number of responses increases with the square of that ratio. Therefore, averaging improves the signal-to-noise ratio by a factor equaling the square root of the number of epochs to be averaged for the various EPs. For instance, a cortical response that has an amplitude of 10 μV and is embedded in EEG activity of 20 μV has a signal-to-noise ratio of 1:2. Averaging only 4 responses improves the ratio by a factor of 2, increasing the signal-to-noise ratio to 1:1. An average of 64 responses increases the signal-to-noise ratio by a factor of 8, making the ratio 4:1. To make the signal 8 times larger than noise, that is, to improve the signal-to-noise ratio by a factor of 16, would require averaging 256 epochs.

Subcortical EPs have a less favorable signal-to-noise ratio and therefore need averaging of larger numbers of responses than cortical EPs. These responses often have amplitudes of only

0.5 μV; if embedded in EEG of about 20 μV, their signal-to-noise ratio is 0.025:1. To bring the EP to the level of the noise, alone, requires a 40-fold improvement, or averaging 1,600 responses. Improvement to 2:1 requires 6,400 responses; 4:1 requires 25,600 responses, an impractical number. This nonlinear relationship between the number of averaged responses and signal-to-noise ratio establishes a practical limitation to the signal-to-noise ratio that can be achieved. This illustrates the importance of good recording methods and of minimizing artifacts during acquisition. Filters can remove much of the unwanted signal, but caution must be exercised because of the effects of filters on the responses, as discussed above. Filtering and averaging can not compensate for sloppy recording methods. If the slow activity on EEG is reduced to 5 μV, a response of 0.5 μV requires only 100 repetitions for a 10-fold improvement in noise ratio to 1:1, and only 1,600 repetitions for an improvement to 4:1.

Long periods of recording an EP should be avoided because they may increase the chance that the signal or noise characteristics vary, for instance, as a result of changes in the subject's level of attention or alertness. Variation of the response characteristics during the averaging will reduce those response parts that vary; such variations can be detected by breaking the average of a large number of responses into smaller parts. Variations of the noise component, for instance, the appearance of slow waves of drowsiness in the EEG, may diminish the signal-to-noise ratio and thereby impair the definition of the EP. Of great practical importance are gross deviations from random noise such as those caused by intermittent large transients, for instance artifacts or K-complexes of the sleep EEG. If such a transient enters the average, it is not reduced by the factor predicted for noise reduction because it does not represent randomly distributed noise. This is why it is important to exclude large artifacts with an artifact rejection option.

The degree of noise reduction is manifested by the amplitude of residual noise in the averaged EP. An estimate of residual noise can be obtained by various methods which may be used to answer the question whether an average contains EP peaks or only noise. Many individuals think of averaging as increasing the amplitude of the signal; conceptually, it is more correct to think of averaging as reducing noise, leaving the stimulus-associated signal of its usual amplitude.

### 3.3.4 Horizontal Parameters of Analysis: Analysis Period, Number of Points, Sampling Rate, and Dwell Time

The choice of the analysis period, epoch, or sweep length, depends on the duration of the EP under study. For transient EPs, this period extends only through the fraction of the interstimulus interval that contains the EP peaks of interest. The most common analysis periods are 200–250 msec for VEPs, 10–15 msec for BAEPs, and 50–100 msec for SEPs. Longer periods are needed to capture long-latency peaks as encountered in EPs of infants, pathologically delayed EPs, and some event-related potentials. For steady-state EPs, the analysis period is made longer than the interstimulus interval so that more than one cycle of the rhythmical response is recorded, usually a few cycles. Because the timing circuits of digital computers operate with binary numbers, the actual duration of analysis is 1.024 times longer than the rounded-off numbers indicated on the settings of most averagers (Table 3.1, columns 1 and 2).

The length of the analysis period is related to the number of points available for averaging and the sampling interval, which is the reciprocal of the sampling rate. The analysis period equals the number of points multiplied by the sampling interval or divided by the sampling rate. In most commercial averagers, the analysis period is selected directly; this selection, in conjunction with the number of points available, determines the sampling interval or sampling rate. However, in a few computers, sampling interval or sampling rate is selected directly, and analysis period is an indirect result of this selection and the number of available points. Most averagers offer 1,024 points per channel. Some older averagers divide the 1,024

TABLE 3.1.   Relationship between parameters of horizontal resolution

| Analysis period (msec) | Sampling rate (kHz) | Sampling interval (msec) | Nyqvist frequency (kHz) | Shortest sine wave resolved (msec) |
|---|---|---|---|---|
| 1,024 | 1 | 1 | 0.5 | 2 |
| 512 | 2 | 0.5 | 1 | 1 |
| 256 | 4 | 0.25 | 2 | 0.5 |
| 204.8 | 5 | 0.2 | 2.5 | 0.4 |
| 102.4 | 10 | 0.1 | 5 | 0.2 |
| 51.2 | 20 | 0.05 | 10 | 0.1 |
| 20.48 | 50 | 0.02 | 25 | 0.04 |
| 10.24 | 100 | 0.01 | 50 | 0.02 |

points by the number of channels, so that with a four-channel recording only 256 points are available per channel, clearly inadequate. All modern machines give adequate numbers of points per channel. Table 3.1 illustrates combinations of the most commonly used analysis periods, numbers of points, sampling intervals, and sampling rates.

In selecting the horizontal parameters of analysis, two requirements must be fulfilled:

• The analysis period must be long enough to include the important peaks of the EP to be studied; and
• The sampling rate or sampling interval must be sufficient to resolve the highest frequencies of peaks.

In particular, care must be taken not to choose a combination that spreads the number of points so widely that they can not adequately depict the high-frequency components of the response under study. To depict a sine wave digitally, at least two points are needed for each cycle. As a minimum, the sampling rate of an averager must be twice as fast as the fastest sine wave component in the signal to be resolved. This critical sampling rate is the *Nyqvist frequency*. Column 4 in Table 3.1 shows the Nyqvist frequency for each sampling rate, and column 5 shows the shortest sine wave that can

be resolved with this sampling rate. The Nyqvist frequency specifies only the minimum frequency for temporal resolution of the components of the EP, but does not ensure the most effective averaging. An increase of sampling rate above this minimum can greatly reduce the number of responses required for a clear definition of the EP; usually at least 10–20 points are used to depict one cycle of the fastest sine wave component of an EP.

The relations between the signal frequency and the sampling rate are illustrated in Figure 3.3, which shows the effect of using the same sampling rate on three sine waves of increasing frequency; whereas waves of relatively low and medium frequency are well resolved in the digital representation, a sine wave slightly faster than the critical value of one-half the sampling rate is not correctly represented after digitization: The frequency of the digitized waveform is lower than that of the original analog sine wave. This erroneous representation of an analog wave by digital values, which suggests a lower frequency due to a sampling rate of less than twice that of the original waveform, is called *aliasing*. Aliasing may be the source of significant errors in EPs and can be minimized by the choice of high-frequency filter settings that eliminate components above the critical sampling rate.

Therefore, when setting up facilities to record a specific EP type, one must determine the frequency of the fastest component of the EP to be recorded and then calculate the limits of the high-frequency resolution resulting from analysis period and number of points to ensure that the highest frequencies in the EP are adequately represented. Although manufacturers of averagers generally tailor their equipment so that this requirement is likely to be fulfilled for the most commonly used EPs, there is no guarantee, and the burden of proof is on the operator. As a minimum, averagers in clinical practice should be capable of a sampling rate of 12.5 kHz, that is, a dwell time of no more than 80 μsec per address; they should have at least 250 addresses per channel. In modern machines, sampling rates of 40 kHz are the minimum and 100 Hz common. The rapid

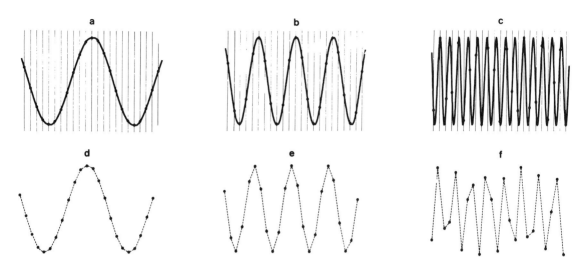

FIGURE 3.3. *The relationship between sampling rate and signal frequency: Effects of different signal frequencies at the same sampling rate. Sine waves at three frequencies (a, b, c) are sampled at the same rate. For each frequency (top traces) the amplitude is measured and represented by a dot at each sampling time. The lower traces are the waves reconstructed by connecting the dots. With a low frequency signal (a) the reconstructed waveform is virtually identical to the native wave. For a faster frequency (b) the reconstructed waveform less well represents the native waveform. For a signal that approaches the sampling frequency (c), the reconstructed waveform does not show all of the peaks, and therefore under-represents the signal frequency. Also, the waveform is distorted, losing the regular appearance of the native signal. Therefore, the sampling frequency must be at least twice the frequency of the fastest signal of interest.*

growth of available memory with newer computers and operating systems makes available memory less of a problem, as well.

### 3.3.5 Vertical Parameters of Analysis: Number of Bits, Resolution, and Recording Gain

At each sampling interval, the digitizer measures the amplitude of the response and converts it into a binary number that consists of binary digits or bits. How well the amplitude of the signal is reproduced digitally depends on the signal size and the number of discrete vertical intervals available for the digital representation. The range of vertical measurements possible is given by the number of bits that the digitizer can assign to the amplitude of the signal at each sampling interval. Such a string is called a *word*.

Most averaging computers have A/D converters with a word length of at least 12 bits; this number of binary digits can represent decimal numbers up to $2^{12}$ or 4,096. Therefore, the converter can resolve 4,096 discrete voltage levels. The voltage resolution can be expressed as the minimum voltage that can be detected by the converter. This is the voltage range of the converter divided by the number of voltage levels. Most converters have different voltage ranges selectable by hardware switches or software programming; for example, −10 V to +10 V, −1 V to +1 V, or 0 V to +5 V, and so on. The −0.1 V to 0.1 V is a common selection. This range is 0.2 V. This dividing 0.2 V by 4,096 voltage levels gives a minimum voltage resolution of 0.049 mV or 49 µV. This would not be acceptable for most evoked potentials, so the raw signal is amplified before it reaches the A/D converter.

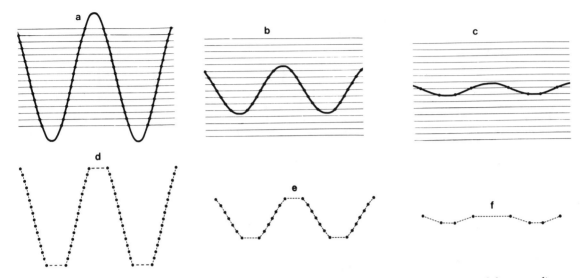

FIGURE 3.4. *The effect of digital conversion on signals of different amplitude. Sine waves of three amplitudes (a, b, c) are digitized using vertical voltage increments of equal size. A dot is marked each time the signal crosses a voltage step. For each amplitude, the top trace is the native signal with the voltage increment overlay; the bottom trace is the waveform reconstructed by connecting the dots of the trace above. With a high voltage signal (a) the reconstructed wave is not able to represent the signal voltage that falls outside the limits of the discrete voltage levels. With a moderate voltage (b) the reconstructed wave is a fair representation of the signal. With a low amplitude signal (c) the voltage resolution is not sufficient to represent the signal accurately.*

The relationship between signal size and vertical resolution is illustrated in Figure 3.4. Signals exceeding the range of the digitizer are distorted; signals occupying about half of the digitized range or more are faithfully depicted in digital form, but smaller signals are inadequately resolved.

For each EP recording, the computer operator must match the signal size with the vertical capacity of the digitizer by adjusting the recording gain. This is easy in averagers that are built so that the vertical range of the digitizer coincides with the height of the oscilloscope display. In these averagers, the gain of the recording amplifier is adjusted so that most responses fill about one half or slightly more of the available display screen. Although an amplitude of 100% of the digitizer would theoretically be better, responses vary so much that selecting a gain that increases the average response amplitude to the maximum capacity of the digitizer results in distortion of many excessively large responses.

Distortions can be avoided if the computer is equipped with an artifact rejection option that can exclude such excessive responses. Lower gain settings require collection of larger numbers of responses; below a minimum gain, even the largest number of responses can not compensate for small amplitude.

Addition and storage of the responses in the computer memory require a wider range of amplitudes than those dealt with by the digitizer; the word length of the memory should exceed that of the digitizer.

## 3.4 AVERAGING DEVICES

### 3.4.1 Artifact Rejection Option

A relatively simple way to prevent some artifacts from entering the average EP is to exclude responses containing deflections of an amplitude that exceeds a selected voltage range. This

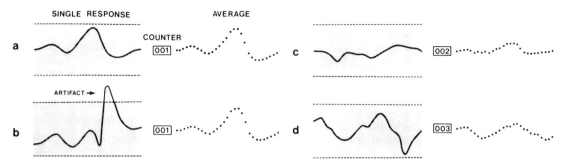

FIGURE 3.5.  *Artifact rejection. A signal must fall completely within voltage boundaries, indicated by the shaded region, in order for it to be accepted into the average* (a). *A signal that does not completely fall within the boundaries is rejected from the average* (b). *Subsequent traces that fall within the boundaries are accepted* (c, d).

procedure is based on the assumption that responses of excessive amplitude are likely to contain artifact. The procedure requires selection of a voltage level likely to include all or most acceptable variations of response amplitude while excluding artifacts larger than those variations; obviously, smaller artifacts will not be rejected and unusually large responses may be rejected. In many averaging computers with built-in amplifiers, an artifact rejection option is coupled with the recording gain so that the computer rejects every response larger than the voltage range displayed on the oscilloscope. Other computers show the artifact rejection option as horizontal lines that mark the allowable amplitude limits of the response. When triggered by a deflection of excessive amplitude, the artifact rejection option prevents the entire response from entering the average (Figure 3.5). An artifact rejection option can be added to computers that do not have one.

### 3.4.2 Oscilloscope Displays

EP computers should allow display of the signal at two or three levels of the acquisition and averaging process (Figure 3.6):

- The average digitized EP is displayed as a sweep that occurs with each response during the averaging and as a steady display after the completion of the averaging (Figure 3.6b).

- Single digitized responses are observed before they are added to the average (Figure 3.6a). In some averagers, this option is chosen with a selector switch as an alternative to the display of the accumulating average; other averagers use a second display sweep to display the individual response while another shows the running average. The single sweep display is used to adjust the windows for the artifact rejection option; it is also useful for identification of artifacts that develop during the recording session.

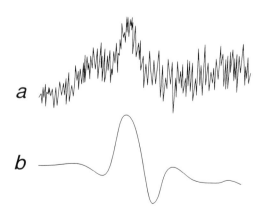

FIGURE 3.6.  *Simultaneous display options during averaging.* (a) *Single response showing the digitized input of a single trial.* (b) *Average of digitized averages.*

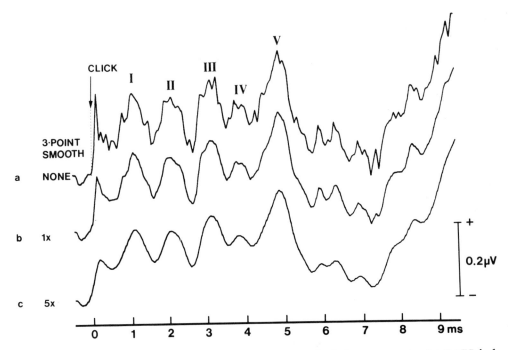

FIGURE 3.7. *Effect of 3-point smoothing on a BAEP containing high-frequency noise. (a) BAEP before smoothing. (b) Same BAEP after smoothing once. (c) Same BAEP after smoothing five times. Note the disappearance of high-frequency components including the stimulus artifact and the better definition of the peaks.*

• Single responses may be observed before digitization and compared with the digitized response to ensure that the digitized version is an accurate representation of the analog signal.

Another feature found in some machines is the option to let the display show a continuous recording of activity at the electrodes. This display may be used to monitor the EEG and to identify artifacts in the recording.

### 3.4.3 Smoothing

High-frequency noise components are often superimposed on EPs and can be reduced by smoothing, a fairly simple digital filtering method (Figure 3.7). This operation replaces each point of a digital tracing by a moving average of the 3, 5, or more neighboring points. The smoothing operation may be used repeat-

edly but must be used with caution. Even though it does not distort the phase relationship between EP peaks, it may change the EP shape by reducing the amplitude of short waves more than that of long ones.

### 3.4.4 Cursors

A cursor is a marker on the display from which the computer makes measurements of latency and amplitude. The shape and function of the cursor differ among machines; a common mode is a vertical line that intersects the trace of interest (Figure 3.8). The cursor can be placed on any point of the digital display. The numerical values of the amplitude and latency of the point indicated by the cursor are shown on the screen. The accuracy of the horizontal readout is limited by the sampling interval; the accuracy of the vertical readout depends on the voltage resolution of the digitizer. Most

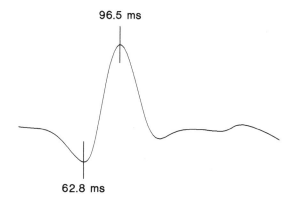

96.5 ms

62.8 ms

FIGURE 3.8. *Function of cursors in EPs. Display details vary among machines, but most show similar information. An averaged trace is displayed and cursors are set at the peaks of interest. Elsewhere on the screen is a table that displays the latencies and amplitudes of the potentials indicated by the cursors.*

averagers provide a selection of one or two or more cursors. If one cursor is selected for a measurement, the readout indicates amplitude and latency of the selected point with reference to the first point of the display or to zero potential. If two cursors are selected, the readout indicates the difference in amplitude and time between the two points marked by the cursors (Figure 3.8); this facilitates measurements of interpeak latencies and of peak-to-peak amplitude. In most modern averagers, changes of the vertical display gain or time base do not alter measurements made with the cursors; the computer compensates for these adjustments. Older machines may not compensate for time base alterations.

## 3.4.5 Addition, Subtraction, and Inversion of EPs

Most averagers allow substantial data manipulation. Recordings may be added to each other, subtracted from each other, and inverted. Addition of averaged responses can help to produce a grand average composed of a larger number of responses. Subtraction and inversion are used to separate waves that differ depending on stimulus phase, for example, to differentiate the cochlear microphonic potential from the potential produced by the action potential of the acoustic nerve.

## 3.4.6 Records of EPs

The EPs are recorded on paper, film, or digital media. This not only provides a permanent record, but is also used for measurements on older machines. There are merits to each format. In general, a hard copy should always be made of each trace used for clinical interpretation; a digital record should not be the only one, because of the risk of data corruption with time and heat. Digital storage is helpful for post-acquisition analysis, comparison with previous recordings, and statistical comparison of data from multiple patients.

### 3.4.6.1 Hard Copies

*3.4.6.1.1 XY plots.* EPs can be plotted in analog form, as a continuous line, or digitally, point-for-point. Digital plotting allows inclusion of letters and numbers such as those used for cursor readouts and labeling coordinates.

The XY plot is usually larger than a photograph, giving better resolution, and plotting paper is relatively cheap. However, XY plotting is slow and may be noisy.

*3.4.6.1.2 Photographs.* Tracings on the face of the oscilloscope can be photographed with a Polaroid camera; the photograph may include an illuminated grid and cursor readouts. Measurements from photographs are accurate only if the horizontal and vertical amplifiers of the oscilloscope are perfectly linear; an illuminated grid on the oscilloscope can be used as a reference only if it is flush with the display and therefore avoids parallactic distortion.

*3.4.6.1.3 Photosensitive paper records.* Older machines make records on photosensitive paper, which must be kept in the dark until it is exposed to a display that reproduces the trace shown on the main oscilloscope screen. Usually calibration marks are also included. The image "develops" seconds after emerging from the machine. This method is not used by modern machines. A related technique is thermal transfer, which produces a recording by heating elements on the print head. This technique allows fast and quiet printing, and is still in use in some newer machines.

### 3.4.6.1.4 Laser printer paper records.
Laser printers have a resolution of at least 300 dots per inch, certainly adequate for EP records. The software usually generates a report that includes not only the traces, but also stimulus and recording parameters, and cursor readouts. Most systems allow the operator to enter an impression.

The mechanism of laser printers is similar to that of copiers. A beam of light shines onto the paper while it is coated with a thin layer of toner. The toner adheres to the paper where the light has hit. Subsequently, the excess toner is recycled into the reservoir.

### 3.4.6.1.5 Bubble-jet paper records.
Bubble-jet printers are used more and more because of their low price, low maintenance costs, and small size. Most understand the PCL5 graphics language used to communicate with laser printers, so they can be used interchangeably. The resolution is similar to that of the laser printers. The only disadvantage is slowness, but this is not important for EP records.

The bubble jet uses a print head composed of linear arrays of microscopic holes. A chamber behind the holes is filled with ink from a reservoir. Tiny heating elements behind the chamber heat the ink until a small bubble forms. This forces ink through the hole so that it impacts the paper, hence the names "bubble jet" or "inkjet." When the element cools, the bubble collapses and ink is sucked into the chamber from the reservoir. The entire process takes about a microsecond.

### 3.4.6.2 Digital records

EPs are usually stored digitally, at least temporarily. The medium may be hard disk, floppy disk, optical disk, or magnetic tape. Digital recordings allow postprocessing of the data, modifying gain, digital filtering, and other manipulation of data.

Digital recording is performed by writing the stream of digitized data in binary format onto the medium. Binary means that the data is represented as a string of 1's and 0's. The length of each block of binary information is called a *word*. For example, if the word length is 8 bits, the number 1 is represented as 00000001; the number 37 is 00100101. The largest number that can be represented by the 8-bit word is 11111111, which is equal to 255. In most modern machines, the A/D converter uses a 16-bit word. This can be represented as a string of 1's and 0's 16 characters long. The largest number that can be represented by a 16-bit word is 65,535.

We can simplify the 16-bit word by making it 4 characters long, adding up every four bits. For example, the 16-bit word 0010000101001001 is equal to 8,521. We group the bits into groups of 4 characters: 0010 0001 0100 1001. Then, decode each group into a decimal number: 2, 1, 4, 9. Therefore, we represent the 16-bit word by the 4-character code 2149. Unfortunately, each character could be greater than 9, for example, if the last group was 1101 or 13. Rather than writing 21413, and creating obvious confusion, we represent the numbers greater than 9 with the letters A to F, so that the word is 214D. This is representing the decimal number 8,525 by the base 16 number 214D. The largest number would be FFFF. When you use a utility program to decode a binary data file, you will often see this type of base 16 or hexadecimal code.

The binary data file is written to magnetic media by the action of a small electromagnet, producing regions of polarization in opposite directions on the media. Optical media are written by laser beams, imprinting the data for 1's and 0's on the disk. The location of the file on the disk is written in the file allocation table (FAT) at the beginning of the disk. Most disks contain two copies of the table, in case one is corrupted. When data is erased from the disk, the location is removed from the FAT directory; the data on the disk is not actually erased, unless the media is "washed." Washing involves intentionally writing nonsense over the data, making retrieval of the original information impossible.

## 3.5 MAGNETIC TAPE RECORDING

The EP raw signal can be recorded to magnetic tape for playback at a later time. This can allow different averaging settings, alterations in the artifact rejection option, and other data manipulations. This differs from digital recording in that it is an analog recording, representing continuous voltage levels rather than a digital data stream. Also, there is minimal preprocessing. With digitally recorded information, only the processed, averaged waves are recorded, so that re-averaging is not possible. In practice, magnetic tape recording is not often performed; few laboratories retain this capability.

## 3.6 OTHER METHODS: FREQUENCY AND POWER SPECTRAL ANALYSIS

Electric responses of the nervous system to sensory stimulation can be analyzed with methods other than averaging. Steady-state responses can be separated from the EEG with a Fourier analyzer or a filter sharply tuned to the stimulus frequency. These methods measure response components having the same frequency as the stimulus. Another method uses power spectral analysis to search for response components defined by the sine-wave frequencies of these components. Both frequency analysis and power spectral analysis yield measures representing amplitude of the response; the clinically more important latency is not measured directly.

# 4

# General Characteristics
# of Evoked Potentials

## 4.1 GENERAL DESCRIPTION OF EPS

EPs consist of a series of peaks or waves in response to a stimulus. Each wave has the following characteristics:

- positive or negative polarity,
- sequential number of the wave,
- latency,
- amplitude, and
- waveshape.

### 4.1.1 Polarity

Polarity refers to positivity and negativity between the two electrodes connected to the inputs of the recording system. The relationship between the electric potential changes at the electrodes and the upward and downward deflections of an EP tracing depend on (1) which electrode is connected to input 1 and which to input 2, and (2) which polarity convention is used in the recording system. Therefore, upward and downward deflections in an EP must be explained in both terms.

### 4.1.2 Number in Sequence

Most EPs have more than one peak. Peaks are labeled differently for each EP type. In some EP types, the number of a peak in the sequence of peaks is used to name the peak. Other EP types are named by combinations of polarity and number in sequence or of polarity and latency, for example, P100 for positive wave of about 100 msec latency in the VEP.

### 4.1.3 Latency

#### 4.1.3.1 Peak identification

Often peaks are not outlined clearly but are obscured by noise. The problem of having to identify peaks in a noisy tracing can often be prevented by reducing the sources of artifacts, such as interference and muscle contraction, by using filters that eliminate high- and low-frequency components outside the range of frequencies represented in an EP, and by collecting a large number of responses. However, in many instances, the problem cannot be eliminated by these means. The tracing then may

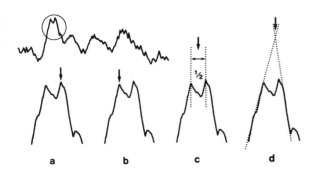

FIGURE 4.1.  *Four methods of selecting the latency and amplitude of an EP peak that is split.* (a) *Highest point.* (b) *First point.* (c) *Latency to the midpoint between two tips.* (d) *Latency to a point extrapolated from the rising and falling phases.*

show abundant noise and raises the questions of (1) how to determine the point at which latency and amplitude of a peak should be measured, and (2) whether there is any EP peak at all.

If two or more small deflections of equal or similar amplitude are found riding on a larger EP wave, the point to be used for measurement of latency and amplitude may be chosen as the highest or the first of the deflections; instead, latency may be measured to a midpoint between the smaller deflections or to the point of intersection of two lines drawn through the ascending and descending portions of the wave on which the smaller deflections ride (Figure 4.1).

Many modern computers use software routines to identify peaks, measure latency, and calculate amplitude. The accuracy of these programs is variable; identification is most accurate if the response is a stereotypic evoked potential. A two-peaked, or *bifid,* evoked potential frequently is mishandled by the automated peak identification; one of the manual methods should be used.

The question of whether a noisy average contains any EP peak often arises when stimuli of threshold intensity are used. An answer may be obtained by comparing the EP to weak stimuli with an EP to stronger stimulus that shows a clearly defined peak and thereby indicates where to look for that peak in the EP to weak stimuli. Another method of detecting peaks in noisy recordings compares the EP in question with an average showing only the noise present during the averaging of the response. An estimate of this noise may be obtained by

- collecting an average of recordings made without stimulation,
- including a period before the stimulus in the average of the responses, or
- averaging the responses while alternating the polarity of the recording to eliminate stimulus-related components, termed *plus-minus averaging.*

These methods are used in some of the peak-measuring computer programs mentioned above.

### 4.1.3.2 Latency of transient EPs

For transient EPs, latency is usually expressed as either peak latency or as interpeak latency. Latency to the onset of an EP deflection is used only rarely because the onset is usually difficult to identify precisely.

*4.1.3.2.1 Peak latency.*  The latency to a peak is measured from stimulus onset to the point of maximum amplitude of a negative or positive peak (LN1, LP1, LN2, in Figure 4.2). The stimulus onset, usually selected as the trigger for the sweep, can be identified easily in most instances because most clinically used stimuli begin suddenly. However, gradually increasing stimuli, such as tone bursts of initially increasing intensity, requires specific definitions of stimulus onset, for instance, the 50% amplitude point or the hearing threshold.

*4.1.3.2.2 Interpeak latency and central conduction time.*  Interpeak latencies (IPLs) are mea-

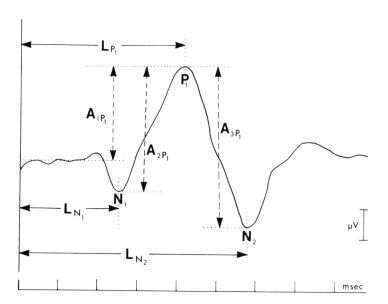

FIGURE 4.2. *Measurement of EP parameters. Peak latency is measured from stimulus to first negative ($L_{N\,1}$), first positive ($L_{P\,1}$), and second negative ($L_{N\,2}$) peak. Peak amplitude is measured from P1 to a preresponse baseline ($A_{1_{P\,1}}$), peak-to-peak of preceding polarity ($A_{2_{P\,1}}$), or peak-to-peak of following polarity ($A_{3_{P\,1}}$).*

sured between peaks of the same EP (Figure 4.3) and are commonly used in the interpretation of BAEPs. They represent the time of conduction between the structures that generate the peaks. The name *central conduction time* is sometimes used instead of *interpeak latency* and is often used in SEP recordings to denote the difference between the latencies recorded from the lower brainstem or spinal cord and the brain. Interpretation on the basis of IPL is preferable to measurement of absolute peak latency because it is less variable and relatively independent of peripheral conduction difficulties. Peak latency

is increased in both central and peripheral conduction abnormalities, whereas IPL is increased by central lesions but not peripheral ones.

### 4.1.3.3 Latency of steady-state EPs

Peak latency of steady-state EPs can not be measured reliably because each peak of these rhythmical responses is the composite result of more than one preceding stimulus; furthermore, the time interval between successive peaks and the time interval between each stimulus and the next peak varies with stimulus

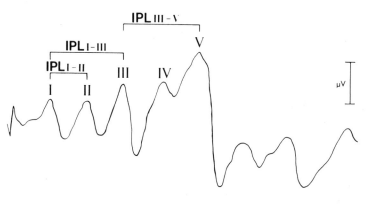

FIGURE 4.3. *Measurement of interpeak latencies (IPL) of the BAEP. IPL I–III is equal to the difference in absolute latencies of waves I and III, but can be measured directly from the tracings. The most important IPLs are shown in the diagram.*

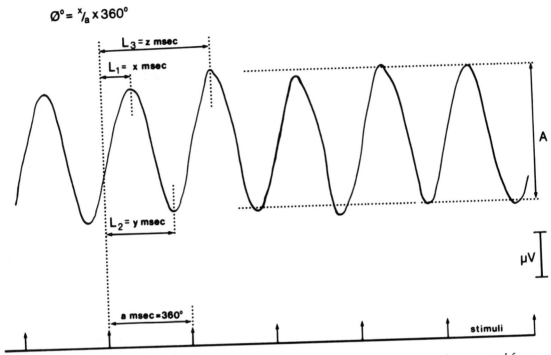

FIGURE 4.4. *Measurement of latency and amplitude of steady-state EPs. Latency may be measured from stimulus onset to the first peak of one polarity, to the first peak of opposite polarity, or to a subsequent peak of defined polarity. Alternatively, latency may be described by the phase angle of the stimulus with respect to the response, that is, if an interstimulus interval is defined as one cycle (360°) the phase angle is the latency from stimulus to response peak divided by the interstimulus interval multiplied by 360°. Amplitude is measured from peak negativity to peak positivity.*

frequency. Several methods can be used to characterize the time relationship between stimuli and response peaks (Figure 4.4):

- Latency may be arbitrarily defined as the interval between a stimulus and the next upward (L1) or downward (L2) peak; a later peak may be chosen for measurements (L3), especially if its latency and polarity resemble those of the major peak of the same polarity of the transient EP.
- The timing may be described by considering the interval between stimuli as a cycle of 360° and calculating the interval between stimulus and response peak in terms of the phase angle between stimulus and peak. For instance, peaks occurring 75 msec after stimuli given

at 10/sec, that is, at 100-msec intervals, may be said to lag behind the stimulus by three-fourths of a full cycle, or by 270°. From measurements of phase lag, an apparent latency of the EP can be calculated.
- Other methods of describing the latency of steady-state EPs have been used but are not discussed here.

### 4.1.4 Amplitude

The amplitude of EP peaks is measured most precisely with cursors. Various types of measurements may be selected for different purposes, but only one type should be used once the selection has been made. Amplitude is clinically a less useful measurement than latency.

### 4.1.4.1 Absolute amplitude measurements

*4.1.4.1.1 Peak amplitude.* The peak amplitude is the vertical distance, representing the voltage difference, between a peak and a reference level representing zero amplitude. This zero level may be defined as a horizontal line drawn through a brief prestimulus interval of the EP recording, through an early poststimulus period not containing EP peaks, or through the average level of the entire EP (Figure 4.2). Measurements of peak amplitude are useful in cases of EPs with a stable baseline and with clearly separated peaks, but these measurements present problems if the baseline is noisy or if a peak begins before the end of the preceding one.

*4.1.4.1.2 Peak-to-peak amplitude.* The peak-to-peak amplitude is the vertical distance between successive peaks of opposite polarity and may be measured as the amplitude of the ascending limb, descending limb, or as a mean of the two (Figure 4.2). Whichever method is chosen, it must be used consistently. Measurements of peak-to-peak amplitude are useful in EPs having stable sequences of peaks of opposite polarity but these measurements present problems when successive peaks vary seemingly independently of each other, or when a peak of high amplitude is preceded or followed by a smaller peak of the same polarity before the tracing returns to baseline. Peak-to-peak amplitude is an important measurement of steady-state EPs (Figure 4.4).

### 4.1.4.2 Amplitude ratios

The ratio of the amplitude of two peaks may be calculated for peaks in the same EP or for peaks that are recorded simultaneously from different locations, such as the two sides of the head or successively from midline and lateral electrode positions.

## 4.1.5 Waveshape

The form of EPs may be fairly characteristic for an EP type and may be described by a figure, for example, W-shaped. However, waveshape varies considerably among patients in the absence of pathology, so shape is not used for interpretation as long as the waves are identified.

## 4.1.6 Potential Field

The appearance of EPs over different parts of the head depends not only on the sensory modality but also on the stimulus used. In clinical practice, the potential field is not used for interpretation, although this may change in the future.

The earliest peaks of cortical EPs are restricted to the specific sensory receiving areas of the stimulus modality. Subsequent peaks have a wider distribution. The two types of peaks are sometimes referred to as *primary* and *secondary*.

## 4.1.7 Naming Peaks

### 4.1.7.1 Nomenclatures

Unfortunately, there is no uniform naming pattern for EP peaks; the nomenclature differs with stimulus modality (e.g., P100 for the VEP, peaks I through VII for the BAEP, and P27, N13, etc. for SEP). The guidelines established by the American EEG Society have standardized the naming. Conventions are presented in the sections on individual EP modalities in this book.

*4.1.7.1.1 Polarity-latency.* Peaks may be named by a combination of the letters *P* or *N*, standing for positive or negative polarity of the peak, and by a number representing the expected peak latency. This method is used for VEPs and SEPs but not for BAEPs. For example, the major peak used for interpretation of the VEP is positive wave at about 100 msec, termed *P100*.

*4.1.7.1.2 Number of peaks in sequence.* Peaks may be named by numbering them in order of occurrence. This method is used mainly for BAEPs whose peaks are labeled I through VII. Because the polarity of the peak is not indicated, it must be known by convention; thus, only vertex-positive peaks are labeled by the numbers of the BAEP. Terminological problems arise if peaks are missing and the numbering of subsequent peaks is in doubt.

### 4.1.7.1.3 Polarity-number in sequence.

Positive and negative peaks may be named by their letters *P* and *N* and by a number indicating their order of appearance among peaks of the same polarity. For example, some investigators call the major positive peak of the VEP *P1*, and the two negative peaks *N1* and *N2*. This method is not in common use.

### 4.1.7.2 Observed versus characteristic peaks

Each nomenclature may be used to describe either observed or characteristic peaks. For instance, a normal patient may have a major positive peak on VEP at 100 msec, so the P100 is properly named. However, a patient with a history of optic neuritis may have a major positivity at 140 msec. Presumably, this is the same peak as the P100 of the normal patient, just delayed by 40 msec. Most laboratories would term this the P100 regardless of the latency, but some use the term *P140*, or *observed latency*. This nomenclature is not in widespread use.

It has been suggested that a horizontal bar be placed over the label of a characteristic peak to distinguish it from the label of an observed peak. For example, a $\overline{P100}$ at 140 msec may be said to be $\overline{P140}$.

A similar problem is encountered when the entire complement of waves is not seen, for example, the BAEP may show only the wave V without the preceding waves. If the latency and other characteristics of the wave confirm identification, the wave is still termed V despite the lack of identifiable predecessors.

## 4.2 VARIABILITY OF EPs

### 4.2.1 Intraindividual and Interindividual Variability

EPs recorded sequentially from the same subject without change in stimulus or recording conditions may differ from each other. EPs recorded with identical methods from different subjects may differ even more. The intraindividual and interindividual variability is very important for clinical diagnostic EP work because the ability to detect abnormalities decreases with increasing variability: The narrower the range of normal EPs, the greater the chance of detecting even slight abnormalities.

### 4.2.1.1 Intraindividual variability

Several factors can alter the ongoing EEG activity, the responses to sensory stimuli, or both. Variation of background activity, or noise, has been held responsible for most of the intraindividual variability of EPs. However, the response, or signal, may also vary during the averaging of an EP.

### 4.2.1.1.1 Changes of attention and alertness.

Attention and habituation may alter EPs, especially the late cortical peaks that are unimportant for most clinical studies. The same is true for sleep. The clinically important BAEPs and SEPs, having relatively short latencies, may be recorded during sleep, which reduces muscle and movement artifacts. Pattern VEPs require focusing on the stimulus and must be recorded when the patient is awake.

### 4.2.1.1.2 Background activity.

VEPs vary with amplitude and phase of alpha rhythm. BAEPs and SEPs are less sensitive to changes in background.

### 4.2.1.1.3 Heart rate and blood pressure.

EPs vary with heart rate and blood pressure; this is usually unimportant except in very small EPs. To avoid distortion of the EP by the heart beat, the stimulus may be time-locked to the cardiac cycle by triggering each stimulus at a constant interval after heartbeats.

### 4.2.1.1.4 Time between averages.

EPs recorded from one subject in different sessions vary more than EPs recorded from the same subject during the same session.

### 4.2.1.1.5 Age.

EPs change fairly rapidly during infancy and then gradually mature to adult levels; most EPs show slight changes during adult age. The variation of EPs among subjects of the same age is greater in infants and children than in adults.

*4.2.1.1.6 Differences among laboratories.* EPs recorded from the same subject in different laboratories with different equipment vary more than EPs recorded from the same subject at different times in the same laboratory.

*4.2.1.1.7 Abnormal cerebral function.* EPs vary more over a diseased cerebral hemisphere than over an intact cerebral hemisphere.

### 4.2.1.2 Interindividual variability

Variations in EPs among individuals are much greater than intraindividual variation because of multiple factors, including sex, body size and proportions, variations in sensory receptors, and central and peripheral neuroanatomy. Particularly important are the spatial orientations of the structures that generate the potential fields that comprise the EP. Differences in skull and skin thickness, resistance, and capacitance are also possible factors. Studies of twins suggest that some variability may be inherited.

## 4.3 SELECTION OF EP TYPE AND OF EP PARAMETERS TO BE MEASURED

### 4.3.1 Selection of EP Type

The most commonly used clinical EPs are VEPs to checkerboard pattern alternation, BAEPs to clicks, and SEPs to electric stimulation of the median and the common peroneal or posterior tibial nerve. The most important parameters are summarized in Tables 7.1, 12.1, 16.1, and 18.1. Transient EPs are used in all three modalities because the latencies of these EPs can be measured easily and are most suitable for detection of abnormal function.

VEPs to checkerboard reversal are less variable and therefore diagnostically more useful than VEPs to diffuse light flashes. Cortical EPs, although more variable than subcortical EPs, must be used for visual studies because sub-cortical VEPs can not be recorded reliably with present methods. Cortical VEPs are recorded with near-field recording methods. They have a long latency and require longer interstimulus intervals than subcortical EPs, but they have a larger amplitude and therefore require collection of fewer responses for an average. Sweep and sampling times are relatively long, and filter settings favor low frequencies.

Among AEPs, the BAEP to click stimuli has become the favorite EP for the diagnosis both of brainstem lesions and of hearing defects in infants. The BAEP is recorded with far-field recording methods. Because it has short latency and duration, it can be elicited with relatively rapid stimulation and recorded with a short sampling interval and with filter settings favoring high frequencies. Because the responses have low amplitude, a few thousand of them must be collected.

SEPs are elicited by stimulation of a sensory or mixed nerve of an arm or leg. SEPs to arm stimulation are simultaneously recorded from scalp, neck, and brachial plexus; SEPs to leg stimulation are recorded from scalp and thoracolumbar spine. These recordings show the ascension of the volley at different points in the afferent pathway. Because SEPs at the different recording sites have different latencies and durations, the parameters of stimulation, recording, and computing are a compromise between those used for far-field and near-field EPs.

### 4.3.2 Selection of EP Parameters to Be Measured

Some EP peaks are less susceptible to variation and more sensitive to pathology than others. In general, latency is the most important measurement used in detecting EP abnormalities. The diagnostic usefulness of EPs having more than one peak or being recorded at different levels is enhanced by measuring IPLs rather than peak latencies because IPLs are less variable and more sensitive to pathology than peak latencies; furthermore, they can distinguish central from peripheral lesions.

Amplitude, showing great normal variability, is much less useful than latency. However, the variability of amplitude measurements may be reduced by measuring peak-to-peak amplitude and calculating amplitude ratios. Peak-to-peak amplitude must be used in the evaluation of steady-state EPs.

## 4.4 LABELING EP RECORDS

Because EPs depend on a great number of variables, the specific conditions for each EP must be documented on the EP record. The guidelines for clinical evoked-potential studies of the American EEG Society suggest that the following items be included:

- subject's name, age, sex, and identification number;
- date of examination and procedure number;
- identification of technologist;
- identification of the electrodes connected to inputs 1 and 2 for each channel;
- the type, intensity, rate, and when relevant, polarity of the stimulus, and the side and site of the stimulus;

- description of other conditions important for the results, such as masking noise delivered to the nonstimulated ear, subject's state of retinal adaptation, level of alertness, degree of cooperation, and apparent fixation on visual stimuli;
- bandpass, and filter roll-off slopes (dB/octave) for the entire recording system, including amplifier, averager, magnetic tape recorder, and any other equipment used in the recording;
- number of responses averaged in each EP;
- horizontal resolution, usually specified as sampling rate for each channel;
- time calibration, usually including an indication of stimulus onset and offset;
- voltage calibration;
- polarity convention indicated by plus and minus signs; and
- marks that indicate which peaks were identified and what latency and amplitude measurements were made.

# 5

# Clinical Interpretation

## 5.1 PRINCIPLES OF CLINICAL INTERPRETATION

In clinical practice, EPs are used to evaluate conduction in the three major sensory systems. Lesions involving these systems often delay, reduce, or abolish EPs; EPs may be abnormal even when clinical examination and other tests are normal. In many instances, the EP abnormalities indicate the location of the lesions. Localization to certain parts of the sensory system may be made either by correlating abnormalities of specific EP peaks with anatomical structures known to generate these peaks or by strategies of unilateral stimulation and recording that take advantage of the anatomical characteristics of crossed and uncrossed portions of the sensory systems.

EPs are less often used to study the peripheral portion of sensory systems, although they may be used to evaluate hearing or visual ability in infants and other noncommunicating subjects.

## 5.2 DEFINITION OF NORMAL: CONTROL GROUPS

Because normal EPs vary in and among subjects, normal values are established by studying normal subjects in several age groups. In general, age-specific normal values should be obtained for each week of age in the neonatal period, for each month in young infants, for each quarter for older infants, and for decades in children and older adults. The specific requirements for age groups vary with EP types. The two sexes should be represented equally in each group of normals. A few laboratories use normative values matched both for age and for sex for EPs showing slight differences between sexes, especially for the BAEP. The normal range of EP characteristics is defined statistically for each subject group, EP type, and EP characteristic. Each group should be composed of at least twenty subjects. Measurements of EPs to stimulation of the right and left eye, ear, and peripheral nerves should not be lumped together unless they show no or negative correlation; however, paired values are useful for the determination of the normal range of intraocular, interaural, and lateral somatosensory differences of latency and amplitude.

Because EPs depend so much on the details of stimulating and recording conditions, control values for each response type to be used in a clinical laboratory must be obtained in that laboratory and can not be adopted from other laboratories. The guidelines of the American EEG Society state that in setting up a new laboratory it is acceptable to use the detailed normative values published by some large laboratories, provided two requirements are fulfilled:

- stimulus, recording, and other conditions in the new laboratory are identical or fully compatible with those of the reference laboratory, and
- a study in the new laboratory of at least twenty normal subjects spanning the age range of the patients to be examined in this laboratory shows that a specified portion, such as 95% or 99%, of these normal values falls within the limits derived from the subject studies in the reference laboratory.

### 5.2.1 Characteristics of Normal Subjects

Normative values are obtained from subjects who are in good general health, free of CNS disease, not under the influence of drugs affecting the CNS, and without a history or findings of neurologic diseases and of disorders of specific sensory modalities under study. Subjects serving as normal controls in VEP studies should have an ophthalmological examination including:

- testing of visual acuity showing a corrected refractive error of no more than −5 diopters for myopes,
- studies of visual fields and color sensitivity, and
- funduscopic examination.

Normal controls for AEPs should have an audiometric examination including pure tone audiometry with threshold measurements for air and bone conduction and tests of discrimination, acoustic impedance, and acoustic reflex. Controls for SEPs should be evaluated with special care if they have a history of sensory defects, trauma, or fractures. For studies making comparisons between scalp recordings from the two sides, the subject's handedness and self-perceived eye or ear dominance should be specified.

### 5.2.2 Statistical Definition of Normal Ranges

The normal range of each EP measurement is defined depending on the distribution of the values of this measurement in normal subjects. Because a statistically normal distribution allows convenient definition of the normal range in terms of mean and standard deviation, an attempt should be made to determine whether the control values for the various measurements are distributed normally, that is, whether a plot of the incidence of each value versus its distance from the mean approximates a bell-shaped curve. Normal distribution may be assessed further by plotting the control measurements on probability paper or, more stringently, by evaluating their fit with the normal distribution curve and by calculating their deviation from symmetry, or skewness, and their degree of peaking, or kurtosis. If the distribution of the measurements is not normal, an attempt should be made to transform them to a normal distribution by such means as calculating their logarithms, square roots, or reciprocals.

After the necessary number of normally distributed groups for each measurement have been determined and a sufficient number of normal values in each group have been collected, the normal range for this measurement may be defined in each group as the mean ±2, 2.5, or 3 standard deviations. This is the normal range for a control sample that represents only a very small part of the normal population. The statistical significance of a measurement that is abnormal with reference to this range can be illustrated by calculation of the tolerance limits which indicate the chance that a certain percentage of the normal population will show measurements falling into the specified normal range. Because in clinical EP studies only excessively large latency values and excessively small amplitude values are considered abnormal, one-tailed tolerance limits may be used.

If the control values for a measurement are not distributed normally, one may plot their cumulative frequency in the control group and use the percentile method to indicate the portion of the normal group having measurements similar to that of an observed EP. To obtain reliable results with this method, the control group should comprise at least 100 subjects. One may use the percentile method to set tolerance limits which exclude 95% or 99% of the normal population, meaning that an observation outside these limits has a 5% or 1% chance of being normal.

Every laboratory must decide on how to set the limits for its normal ranges. These limits represent a compromise between the numbers of false-positive and false-negative readings that can be accepted. Choosing a wider normal range has the advantage of decreasing the number of false-positive readings but also has the disadvantage of increasing the number of false-negative readings. Where best to draw the line between normal and abnormal values is a practical matter that is ultimately determined by the ability of the test results to discriminate healthy from diseased subjects.

A simple graphic method may be used to compare latency and amplitude measurements of an EP peak with normal values. The normal values for this peak may be plotted against each other, showing latency on the horizontal axis and amplitude on the vertical axis. The normal range can then be drawn in the form of an ellipse containing a defined percentage of all normal values; abnormal values fall outside this ellipse of probability.

## 5.3 TYPES AND CLINICAL IMPLICATIONS OF EP ABNORMALITIES

### 5.3.1 Abnormally Long Latency

Increased latency is much more reliable than decreased amplitude in signalling clinical abnor-malities in a sensory system. Increased latency generally indicates decreased conduction velocity and is often due to lesions that cause slowing of axonal conduction in the central part of the system. However, peripheral lesions can also increase the conduction time. The latency of VEPs to patterned stimuli may be increased as a result of ocular conditions that blur the retinal image; BAEPs may be delayed by cochlear lesions and SEPs may be delayed due to impaired peripheral nerve condition, Therefore, before an increased latency can be interpreted as indicating a lesion in the central part of a sensory pathway, peripheral lesions must be excluded. Measurements of interpeak latencies may help distinguish central from peripheral lesions.

A very common cause of slowed conduction is the demyelination of nerve fibers, which reduces conduction velocity both by delaying saltatory conduction and by replacing it with very slow continuous conduction. These factors can account for the delays of EP peaks sometimes seen in traumatic, ischemic, degenerative, and other demyelinating lesions. Reduced conduction velocity contributes to the very long delays characteristic of multiple sclerosis. The resulting dispersion of impulses probably reduces the efficiencies of temporal and spatial summation of postsynaptic potentials. Restoration of conduction by remyelination is likely to play an important part in the recovery from acute and chronic compressive lesions of the CNS but not in the remissions of multiple sclerosis. This lack of effective remyelination probably explains the rather high incidence of abnormal EPs in patients with stable multiple sclerosis.

### 5.3.1.1 Abnormally long peak latency

Increased peak latency indicates a defect between the point of stimulation and the structure generating the EP peak, that is, the stretch from retina to occipital cortex in VEPs, from cochlea to pons and midbrain in BAEPs, and from peripheral nerve to spinal cord or somatosensory cortex in SEPs. Peak latency is the major parameter of VEPs; it is less important than IPLs in BAEPs and SEPs.

### 5.3.1.2 Abnormally long interpeak latency

Interpeak latencies can be measured in BAEPs and SEPs and are preferred to peak latency for three reasons: (1) The peaks representing acoustic nerve excitation in BAEPs and the peak indicating brachial plexus or cauda equina excitation in SEPs serve as benchmarks that help to distinguish peripheral lesions, which cause a delay of these peaks and all subsequent ones, from central lesions, which leave the latency of the benchmark peaks intact but delay the subsequent ones and thereby increase the IPL or central conduction time; (2) latencies measured between peaks may help to localize lesions to the structure in the central sensory path assumed to generate these peaks; and (3) interpeak latencies vary less than peak latencies and are more sensitive to pathology; they are therefore better indicators of abnormal function.

### 5.3.1.3 Abnormal differences of peak or interpeak latencies

Differences of peak or interpeak latencies between EP peaks recorded from the midline in response to successive stimulations of the eye or ear on either side (intraocular or interaural latency difference) or between EPs recorded simultaneously on both sides of the head (left-right latency difference) may suggest an abnormality of the EP having longer latency. Not all EP measurements identified as abnormal by comparison with the normal range of latency differences fall outside of the normal range of absolute peak or interpeak values.

### 5.3.2 Abnormal Amplitude

A decrease of amplitude often accompanies the increase of latency seen in cases of reduced conduction velocity and is probably explained by dispersion due to general slowing of conduction, increased scatter of conduction velocities, and complete dropout of some of the fibers. Primary axonal defects produce reductions of amplitude rather than increases of latency. In general, amplitude abnormalities are less clinically useful than latency abnormalities.

### 5.3.2.1 Reduced amplitude

Low amplitude has limited value as an indicator of impaired conduction because amplitude varies much more than latency in normal subjects and because amplitude reduction does not necessarily indicate conduction defects but may be due to other defects, including technical problems. Therefore, amplitude reduction can suggest a conduction defect but usually can not prove it.

### 5.3.2.2 Abnormal amplitude ratios

The amplitude ratio of EP peaks is a somewhat better indicator of abnormality than is absolute amplitude because it is less variable; also, an abnormal amplitude ratio is less likely to be due to technical problems. Several kinds of amplitude ratios can be used:

- Amplitude ratios of peaks in the same EP are used for the interpretation of BAEPs; a commonly used measurement is the amplitude ratio of waves V/I, or its reciprocal, which may indicate brainstem lesions.
- Amplitude ratios of corresponding peaks of EPs recorded simultaneously from the left and right side of the head may be used to search for unilateral hemispheric lesions, such retrochiasmal lesions of the visual pathway.
- Amplitude ratios of peaks of EPs recorded successively from the midline are used to evaluate VEPs to stimulation of either eye in the search for a prechiasmal abnormality.
- Amplitude ratios of peaks of EPs recorded successively from the two sides of the head may be used in studies of SEPs to stimulation of peripheral nerves of either side of the body and may reveal unilateral lesions in the somatosensory system.

### 5.3.2.3 Complete absence of an entire EP

Absence of an EP may indicate complete interruption of conduction at a point before that generating the first EP peak. This severest degree of EP abnormality may represent the maximum effect of a lesion which in milder form produces only a decrease of amplitude, an increase of latency, or both. However, complete absence of an EP does not necessarily prove a conduction defect because it may be due to other factors, including peripheral problems such as eye or ear disorders, technical problems such as failure of stimulation or of synchronization between stimulator and averager, faulty electrode connections, or defects of electrodes or data acquisition equipment. Only after these problems have been excluded can the absence of an EP be attributed to a CNS lesion.

### 5.3.3 Abnormal Shape

The waveshape of EPs cannot be easily defined and is not a reliable indicator of abnormality. If the waves are identifiable and of normal latency, the EP is interpreted as normal regardless of shape.

Abnormal shape is usually due to variations in spatial orientation of the neuroanatomical structures responsible for the EP waves.

## 5.4 STRATEGY OF LOCALIZATION AND LATERALIZATION

Chiefly two principles are used to determine the side and level of a lesion.

### 5.4.1 Correlation between EP Peaks and Anatomical Structures

Certain peaks in the BAEP and the SEP correspond with excitation of certain structures in the auditory and somatosensory pathways so that lesions of these structures can be detected by abnormalities of the corresponding peaks; subsequent peaks, representing excitation of more proximal structures that receive abnormal input from the damaged structure, are usually also abnormal. Correlations of EP peaks with anatomical structures have been made empirically and are useful for an approximate localization even though it is now agreed that one-to-one correlations between a peak and a specific point in the sensory pathway can not be made because most peaks are due to activity in more than one

structure and any structure may contribute to more than one peak.

### 5.4.2 Localization by Combinations of Unilateral Stimulation and Recording in Partially or Completely Crossed Pathways

Each of the major sensory pathways can be divided into an uncrossed, crossing, and crossed part, as shown in Figure 5.1. The crossing is partial in the visual and auditory systems, connecting each side of the body to both hemispheres, and complete in the somatosensory system, connecting one side of the body to the opposite hemisphere only. Lesions in these three body parts can be detected by combinations of unilateral stimulation and recording. A lesion in the uncrossed part causes abnormalities in EPs to stimulation of only that side. A lesion in the crossing part produces abnormalities of EPs to stimulation of either side. These abnormalities often differ from each other. A lesion in the crossed part of the somatosensory system causes abnormalities of the SEP to contralateral stimulation. A lesion in the crossed part of the visual system is best searched for by selective excitation of each hemisphere with visual half-field stimuli and by recording bilateral EPs. A lesion in the crossed part of the auditory system is difficult to detect with any strategy.

## 5.5 REPORT

A report should be provided for every clinical EP study. The report should give a description of the findings, including measurements of peak latency, interpeak latency, amplitude ratio, and other criteria used for the definition of normal responses. The report should indicate which measurements are abnormal and what methods were used to define the normal range, for instance, the number of standard deviations beyond the normal mean or the tolerance limits. Relevant details such as visual acuity, hearing threshold, treatment with drugs having possible effects on EPs, level of alertness, and cooperation during the test should be included.

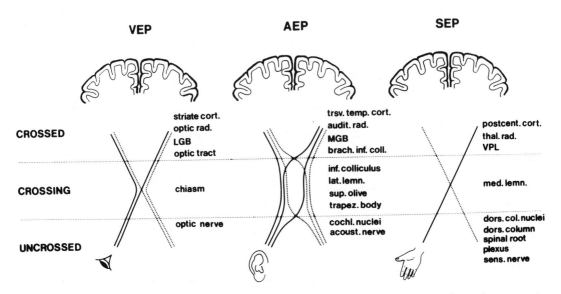

FIGURE 5.1.   *Localization of lesions in central sensory pathways. For each modality, the pathway consists of a distal uncrossed segment, a crossing, and a central crossed segment.*

The report should give a clinical interpretation, especially in cases of abnormal findings. The interpretation should point out the implications of the abnormal findings in general pathophysiological terms and with respect to the clinical data of the specific case under study.

# B

# Visual Evoked Potentials

## PART CONTENTS

# 6

# VEP Types, Principles, and General Methods of Stimulating and Recording

## 6.1 VEP TYPES

Several types of VEPs are used in routine clinical practice and investigation (see Table 6.1). These modalities are distinguished by:

- visual content,
- rate,
- mode of presentation,
- type of stimulator used, and
- stimulus parameters such as stimulus size and brightness.

These aspects are discussed in the sections of this chapter on the specific types of VEPs.

The classification of VEPs, which depends primarily on stimulus characteristics, differs from the classification of AEPs, which depends on recording methods, and of SEPs, which depends on the location of the stimulus and recording electrodes.

## 6.1.1 Visual Content: Patterned and Diffuse Light

The visual content of a stimulus can be divided into two types: patterned and diffuse. The most important patterned stimuli are

- checkerboard patterns consisting of light and dark squares with sharp borders, and
- sine wave grating patterns consisting of light and dark stripes with a gradual transition of brightness between them. Other patterns such as bars with sharp borders, small light spots, and random dots are used occasionally. VEPs to patterned stimuli (pattern VEPs) are due mainly to the visual content of the pattern, especially the density of light and dark contrast borders; VEPs to diffuse light flashes (flash VEPs) are due to changes in luminance only.

TABLE 6.1.  VEP types

A. VEPs to patterned light
  1. VEPs to checkerboard patterns
    a. Transient VEPs
      (1) VEP to checkerboard pattern reversal
          and shift
      (2) VEP to checkerboard pattern appearance
      (3) VEP to checkerboard pattern disappearance
    b. Steady-state VEPs
      (1) VEP to checkerboard pattern reversal
          and shift
      (2) VEP to checkerboard pattern appearance–
          disappearance
      (3) VEP to checkerboard pattern flashes
  2. VEPs to sine wave gratings
    a. Transient VEPs
      (1) VEP to sine wave grating reversals
      (2) VEP to sine wave grating appearance
      (3) VEP to sine wave grating flashes
      (4) VEP to sine wave grating contrast-depth
          modulation
    b. Steady-state VEPs
      (1) VEP to sine wave grating reversals
      (2) VEP to sine wave flashes
  3. VEPs to other patterns: bar gratings, small light
     spots, random dots
B. VEPs to diffuse light
  1. Transient VEPs
    a. VEP to brief flashes of diffuse light
    b. VEP to onset of diffuse light
    c. VEP to end of diffuse light
  2. Steady-state VEPs
    a. VEP to repetitive flashes
    b. VEP to sine wave modulation of light intensity
C. Other VEPs: Scotopic VEPS, VEP to colored light
   stimuli, VEPs to blinking, eye movements, pattern
   motion, electric stimulation of the eye

## 6.1.2 Stimulus Rate: Transient versus Steady-State VEPs

Transient VEPs consist of a sequence of different peaks that occur at a constant latency after each stimulus. Steady-state VEPs consist of a rhythm of uniform peaks occurring at the same frequency as the repetitive stimulus or at harmonic frequencies. In clinical studies, transient VEPs are used much more often than steady-state VEPs.

## 6.1.3 Presentation Mode: Pattern Reversal and Pattern Shift, Pattern Appearance and Disappearance, Patterned and Diffuse Light Flashes

Presentation modes differ for patterned and diffuse light stimuli. In the commonly used method of alternating patterns, the stimulus consists of a sudden change of all light pattern elements into dark ones, and vice versa. This alternation may be accomplished either by replacing each stationary pattern element by an element of the opposite phase (pattern reversal) or by displacing the pattern by the width of one pattern element (pattern shift). The term *pattern reversal* is often used to denote both stimulus mechanisms.

Pattern appearance and disappearance are produced by presenting a pattern alternating with a diffusely illuminated or dark field; either the onset or the end of the pattern presentation is used as a stimulus. Patterned light flashes are pattern presentations lasting so briefly that they elicit a VEP combining the responses to onset and end of the pattern presentation. Diffuse light stimuli are usually presented as brief flashes producing combinations of responses to the onset and end of illumination. Longer light pulses elicit VEPs to either the onset or the end of diffuse light and are not used for routine clinical studies. Only a few of the many possible combinations of patterns and presentation modes are used clinically.

## 6.1.4 VEPs Most Commonly Used in Clinical Practice

In general, transient VEPs are preferred to steady-state VEPs because their normal range is more concisely defined. Pattern VEPs, especially transient VEPs to alternating checkerboard patterns, are generally preferred to flash VEPs for several reasons:

- pattern VEPs vary less among subjects and are more sensitive to lesions that impair conduction through the visual pathway;

- patterned light can be used more conveniently for separate stimulation of the left and right visual half-fields, which is necessary for investigation of chiasmal and retrochiasmal lesions;
- patterned stimuli can be used for ophthalmological testing of visual acuity and refractive errors.

Transient flash VEPs have been used extensively in the past but are now largely abandoned because they vary so much in and among subjects that they are relatively insensitive indicators of abnormalities. However, they are still used in patients who can not focus on a patterned stimulus, especially in infants and small children, comatose patients, and subjects with poor visual acuity. Steady-state VEPs to diffuse light flashes are used in a few laboratories to determine the highest flash rate capable of eliciting a steady-state VEP, or the critical frequency of photic driving. Flash VEPs can not give details of the functioning of the visual system; rather they indicate integrity of the system when present.

VEPs to sine wave gratings have not yet been used much in clinical practice, although their diagnostic value may meet or exceed that of the checkerboard patterns. Sine wave grating stimuli are more difficult to handle, and less information on normal and abnormal responses has been accumulated.

There is no ideal method of eliciting VEPs. The type of VEP to be used depends on the requirements of the case. Although transient VEPs to checkerboard reversal generally have the greatest ability to detect abnormalities, in some patients abnormalities can be detected only with other stimuli. Therefore, a few laboratories now routinely use more than one type of VEP, for instance, a combination of transient VEPs to checkerboard pattern reversal and steady-state VEPs to diffuse light. In some cases, stimulus and recording methods may need to be changed and other VEPs may need to be added for a complete examination.

## 6.2 THE SUBJECT

Subjects should be seated in a comfortable chair in a quiet room and should remain alert during the recording. Use of a chin or neck rest can reduce muscle artifact in the recording. Visual stimuli may be presented on a TV screen, or on a transparent or reflecting screen at a fixed distance from the subject's eyes. Stroboscopic flashes may be applied from a lamp directed at the subject's eyes. The stimulator should not produce any noise that could interfere with the VEP or elicit an AEP.

A steady, dim background illumination is usually maintained to reduce the importance of stray light from various sources and to stabilize dark adaptation during long intervals between stimuli. Pupillary dilatation or an artificial pupil should be used for flash stimulation but is not needed for routine patterned light stimulation, although extreme meiosis, ptosis, and cataracts may reduce the intensity of patterned stimuli enough to affect the VEP. More important for pattern VEPs is visual acuity: Subjects requiring corrective lenses for the stimulus distance should wear their glasses and should acknowledge that they can see the stimulus pattern sharply because blurring of small pattern elements may alter the VEP. Furthermore, without their glasses, some myopic subjects squint to see the target more clearly, and this pinhole effect may reduce the stimulus luminance and thereby change the VEP.

Monocular stimulation is best carried out by covering the nonstimulated eye with an eyepatch. Having the patient close one eye or hold an ocular occluder is less satisfactory because it is likely to introduce muscle and movement artifacts into the recording. A fixation point is usually used with full-field stimulation and must be used with half-field stimulation studies. Responses to the first few stimuli are often not included in the average because they may be affected by startle or head and eye movements. The beginning of averaging should be delayed, especially for steady-state VEPs, which often develop only after an initial transition period of at least a few stimuli.

## 6.3 GENERAL DESCRIPTION OF STIMULUS PARAMETERS

A change of a stimulus parameter can change VEP amplitude and latency. The effect is not easily predictable because it occurs only within a certain range between a minimum, or threshold, and a maximum, or saturation, point; even within this range, the relation between a change of stimulus parameter and changes of VEP may be nonlinear. Furthermore, changes of one parameter interact strongly with other parameters and may shift the entire range of effectiveness of another parameter. As a practical consequence, laboratories doing clinical VEP studies must select a specific type of stimulus defined by visual content, presentation mode, stimulator, and stimulus parameters and strictly adhere to that selection in order to avoid the variations of VEPs that occur with the slightest deviation from routine.

### 6.3.1 Stimulus Rate

For transient VEPs, stimulus rates of 1–2/sec are usually used. Slower stimulus rates unnecessarily prolong the recording session and introduce the risk of increasing VEP variability due to changes in attention. At stimulus rates of about 4–6/sec, individual VEPs begin to interact with each other. A steady-state response of uniform, sinusoidal shape is usually obtained at rates of more than 6–8/sec and may be largest at one or more specific stimulus rates. Flash stimuli more than 50/sec may be needed to determine the critical frequency of photic driving.

### 6.3.2 Intensity

Luminance is the most important parameter for diffuse light stimuli because they act only by virtue of a change of luminance. In contrast, patterned light stimuli are used with the aim of avoiding overall luminance changes; however, changes of luminance affect the VEPs to both types of stimuli.

*Luminance* refers to the amount of light coming from a surface and can be measured by comparing the brightness of the surface with that of a standard light source. The luminance of a patterned light stimulus is defined by indicating the luminance of the light and dark pattern elements, which may be measured with a precision spot photometer. The mean luminance of a pattern equals the mean of its light and dark elements. The luminance of short flashes can be defined by the luminance of the steady light produced by flashes given at a rate above the fusion frequency; however, in most stroboscopic lamps, the light output per flash decreases with increasing flash rate. Because it is difficult to measure the intensity of brief flashes, intensity is often characterized by specifying the control settings and the distance of the lamp from the subject's eye for equipment made by a particular manufacturer.

Luminance is measured as candela per square meter ($cd/m^2$); older measures, which can be converted to candela per square meter, are stilb ($cd/cm^2$), footlambert ($cd/\pi ft^2$), lambert ($cd/\pi cm^2$), and apostilb ($cd/\pi m^2$). Retinal illumination is measured in trolards, one trolard being the illumination that results when a surface with a luminance of 1 $cd/cm^2$ is viewed through an aperture of 1 $mm^2$.

The effect of a light stimulus varies with pupillary size to a degree that depends on the stimulus type. VEPs to patterned stimuli that are not associated with significant luminance changes are not affected by changes in pupillary size unless they are extreme; pupillary size is therefore usually not controlled in recordings of pattern VEPs. In contrast, VEPs to diffuse light, depending entirely on the change of retinal luminance, vary greatly with pupillary size, which therefore must be controlled in recordings of these VEPs to reduce the variability of the responses. The effect of pupillary size can be controlled by cycloplegics, an artificial pupil, or a Maxwellian system (chapter 7). In practice, however, the flash VEPs are not given much weight in clinical interpretation, other than to determine integrity of the visual system.

The effectiveness of a light stimulus depends to some extent on the amount of ambient light present during the recording. Although ambient

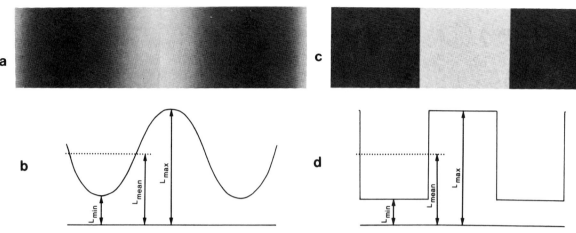

FIGURE 6.1. *Stimulus type for pattern VEPs. Left is a sine wave grating and right is a square wave pattern. For each, the top is the visual appearance and the bottom is a graph of luminance. For each pattern, there is a luminance minimum ($L_{min}$), maximum ($L_{max}$), and mean ($L_{mean}$). These values are shown on the luminance graphs.*

light generally reduces the effect of a visual stimulus, keeping the background of the recording room very dimly lit helps to make stray light from oscilloscope sweeps, illuminated dials, control lamps, computer screens, and other sources insignificant and to prevent dark adaptation, which may develop between stimulations. The ambient luminance can be determined by measuring luminance at several sites near the stimulus field and by calculating the mean of these measurements.

### 6.3.3 Brightness Contrast of Pattern Stimuli

#### 6.3.3.1 Contrast borders: Checkerboard and sine wave grating patterns

Transitions between light and dark pattern elements have sharp borders in checkerboard patterns and in bar gratings. These transitions can be represented by a spatial square wave (Figure 6.1a). In sine wave gratings with sinusoidally modulated brightness contrast, the change of luminance between light and dark stripes is gradual and has the form of a sine wave (Figure 6.1b).

#### 6.3.3.2 Contrast depth: Counterphase and depth modulation stimuli

$$Contrast = \frac{Lmax - Lmin}{Lmax + Lmin}$$

The *depth of contrast* is defined as the difference between the luminances of a light ($L_{max}$) and a dark ($L_{min}$) element divided by their sum. By this definition, maximum contrast has a value of 1, minimum contrast a value of 0. In clinical practice, contrast of stimulus patterns is usually set to at least 0.5, that is, to a ratio of at least 3:1 between maximum and minimum luminances.

Clinical studies use such extreme changes of contrast as produced by pattern reversal, appearance, and disappearance. Research studies have investigated the effect of partial changes of contrast produced by intermittent increase and decrease of the brightness of the light and dark pattern elements. This contrast depth modulation is a less effective stimulus than complete pattern reversal. The timing of the contrast reversal is usually as quick as possible, but a gradual alternation can be produced by stimulators using rotating Polaroid filters: One filter rotates over a fixed filter. Light transmission will

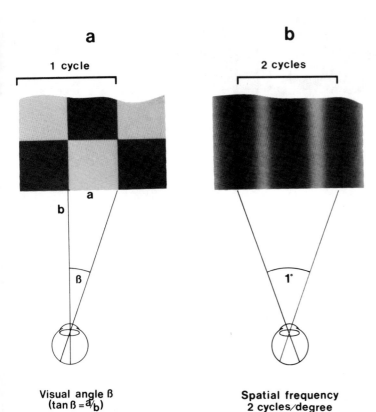

**a**

1 cycle

**b**

2 cycles

Visual angle ß
(tan ß = $a/b$)

Spatial frequency
2 cycles/degree

FIGURE 6.2. *Two methods of describing the size of a pattern visual element: (a) Visual angle ß (tan ß = a/b) is defined as the subtense of a single light or dark element at the retina. (b) Spatial frequency is the number of recurring pattern elements, or cycles, per one degree of visual angle.*

be best when the filters are aligned perfectly (angle of 0°); transmission will be least when one filter is rotated 90°. At angles in between, there is a smooth gradient of light transmission.

### 6.3.4 Size of the Elements of Pattern Stimuli

The bright and dark elements of checkerboard and grating patterns most commonly used in clinical practice have equal size. Figure 6.2 illustrates measurements of visual angle and spatial frequency commonly used to describe this size.

#### 6.3.4.1 Visual angle

The visual angle ß describes the size of the image of one light or dark element at the retina and is derived from tan ß = $a/b$, where $a$ is the side length of the element and $b$ its distance from the eye (Figure 6.2a). Visual angles of less than 1° can be approximately calculated by ß = 3,450 $a/b$ in minutes of arc; angles of 1° or more can be calculated by ß = 57.3 $a/b$ in degrees of arc.

Visual angle is commonly used to indicate the size of the pattern elements with sharp borders such as the squares in checkerboard patterns.

#### 6.3.4.2 Spatial frequency

The spatial frequency equals the number of repetitions of one light plus one dark element, or of one cycle of the stimulus pattern, per degree of visual angle (Figure 6.2b). The elements defined by spatial frequency and visual angle are different: A light or dark element of a visual angle of 15' is only half of the element described by the corresponding spatial frequency; this element, consisting of a light and a dark unit of 15' each, has a visual angle of 30'.

Spatial frequency, rather than visual angle, is commonly used to define the size of the elements in sine wave gratings that represent a single spatial frequency. Spatial frequency is used less often to indicate the size of the checkerboard squares. The spatial square waves representing the light and dark squares of a checkerboard pattern consist of a mixture of frequencies. If the light and dark elements of a checkerboard pattern were to be represented by alternating light and dark stripes, the direction of the stripes would run at right angles not of 90° but of 45° to the squares. The major Fourier component of the spatial frequencies contained in checkerboard patterns is oriented diagonally to the squares and has a wavelength of 1.4 times the side-length of the squares.

### 6.3.5 Size of the Stimulus

Field size is best described in terms of visual angle subtended by the stimulus. Assume that the eye is in the middle of a sphere, which represents the surrounding visual space. The eye can only see a limited amount of this sphere, without movement of the head or eye. Stimulus size refers to the size of this space, in degrees arc, occupied by the stimulus.

### 6.3.6 Full-Field and Partial-Field Stimulation

For full-field stimulation, the stimulus pattern extends equally to both sides of a fixation point, such as a small light-emitting diode, placed in the middle of the pattern. Hemi-field stimulation may be accomplished in two ways: (1) One half of the stimulus pattern is presented to the left or right of the central fixation point; (2) the entire stimulus pattern is used by positioning the fixation point at the right or left margin of the pattern. With either method, the fixation point should be about 0.5–1° outside the stimulated hemi-field to avoid stimulation of the opposite hemi-field due to small involuntary shifts of gaze. In method (1), a dark strip about 1° wide may be placed vertically across the middle of the stimulus pattern, and the fixation point may be positioned in the middle of the strip in the upper or middle part of the pattern.

Stimulation of the upper and lower hemiretina may be accomplished either by presenting only the part of the stimulus pattern above or below the fixation point or by moving the fixation point to the upper or lower border of the entire pattern.

### 6.3.7 Pattern Orientation

Patterns whose components are oriented vertically are more effective than obliquely oriented patterns. The major spatial Fourier component of checkerboard patterns is diagonal to the checks.

## 6.4 GENERAL DESCRIPTION OF RECORDING PARAMETERS

Although stimulus characteristics differ for each type of VEP, most recording parameters are similar for all VEP types.

The number of channels should be four for a complete examination including the retrochiasmal part of the visual pathway, but fewer channels are often used for testing of the prechiasmal part.

Recording electrodes are placed on the occipital scalp in the midline and laterally; reference electrodes are positioned on the frontal or central scalp or on the earlobes. The precise location of these electrodes and the electrode combinations used for recording differ widely in different laboratories and vary for different types of VEPs. Specific recommendations have been made for VEPs to checkerboard pattern stimulation and to diffuse light flashes.

The polarity convention is not standardized; major positive or negative peaks are displayed upward in some laboratories and downward in others.

Guidelines of the American EEG Society recommend the following electrode positions for VEPs:

- MO: Midline occipital, 5 cm above the inion,
- MF: Midline frontal, 12 cm above the nasion,
- RO: Right occipital, 5 cm to the right of MO,
- LO: Left occipital, 5 cm to the left of MO.

The following montage is recommended for pattern-reversal VEPs:

- Channel 1: RO–MF,
- Channel 2: MO–MF,
- Channel 3: LO–MF.

Filters are set to a bandwidth between 0.2–1.0 Hz and 200–300 Hz. A decrease of the high-frequency filter to 100 Hz may cause an apparent increase of peak latency.

The analysis period for transient VEPs is 250 msec in normal adults and up to 500 msec in infants and for abnormally delayed VEPs at any age. The requirements for the temporal resolution of VEP waves during these periods are easily met by conventional averagers: Even in an averager with only 250 points per channel, a sweep length of 500 msec results in a dwell time of 2 msec, or a sampling rate of 500 Hz, which can resolve waves up to 250 Hz, that is, waves much faster than those contained in VEPs. Usually 100–250 responses are averaged and at least two averages are superimposed to ascertain replication.

# 7

# The Normal Transient VEP to Checkerboard Pattern Reversal

## 7.1 DESCRIPTION OF THE NORMAL TRANSIENT PATTERN REVERSAL VEP

The most commonly used methods of producing and analyzing VEPs are summarized in Table 7.1. Figure 7.1 shows a normal midline occipital VEP to checkerboard pattern reversals at 2/sec. It is characterized by a positive occipital peak at a latency of 90–110 msec and an amplitude of about 10 μV. This peak is often referred to as the $P\overline{100}$ or $P1$.

In many cases, this major positive peak is preceded by a smaller negative peak at about 60–80 msec, sometimes called $N\overline{75}$ or $N1$, and followed by a second negative peak, $N\overline{145}$ or $N2$. A second positive peak and later peaks vary considerably in and among subjects and depend on the location of the reference electrode. In some instances, the first positive peak is so closely followed by a second positive peak that the two peaks may seem to be part of a single deflection having two smaller peaks. Bifid peaks differ from the splitting of a single peak produced by superimposed residual noise in that the

separation between the split peaks is deeper and wider than that caused by noise in other parts of the same VEP. The bifid pattern may be due to visual field defects or to a difference in potential fields generated by the lower and upper halves of the visual cortex. In the former case, recording from lateral electrodes may help identify the true $P\overline{100}$. In the latter case, stimulation of the lower visual field may produce a single-peak response, since only the superior aspect of the visual cortex will be generating a response. A very small positive peak occasionally precedes the $P\overline{100}$ (Figure 7.1) and is of no clinical importance, but has led some investigators to use the terms $P1$ for this peak and $P2$ for the major positive peak. This is not conventional terminology. This bifid pattern is so much more often seen in patients with conduction defects than in normal patients that its presence should be considered abnormal.[53]

The precise values of the mean latencies and amplitudes depend on many factors, which are described individually in following sections of this chapter. Most of these factors have similar effects on transient and steady-state VEPs,

TABLE 7.1. VEPs to checkerboard pattern reversal

A. Subject variables
    1. Age: Separate normal control groups are needed for subjects up to about 5 years and over 60 years.
    2. Sex: Separate controls for males and females are used in most laboratories.
    3. Visual acuity: Visual acuity should be better than 20/200 for squares of 1° or less. The subject should see the stimulus pattern clearly. Refractive errors should be corrected.
    4. Pupillary size: Pupils should neither be altered by medication nor fixed at extremely small or large size by disease.
B. Stimulus characteristics
    1. Type: Checkerboard reversal on TV screen, checkerboard pattern shift by pivoting mirror projection
    2. Rate: 1–2/sec
    3. Phase: Responses to reversal into one phase are averaged with responses to reversal into the opposite phase.
    4. Brightness contrast: Greater than 0.5 for bright versus dark squares
    5. Intensity: Luminance of bright and dark squares and ambient luminance must be kept constant.
    6. Monocular stimulation: Monocular stimuli should be used for full-field and hemi-field stimulation.
    7. Full-field versus hemi-field stimulation: The stimulus pattern extends equally to the sides of the fixation point for full-field stimulation, but only to one side for hemi-field stimulation.
    8. Check size: 28–31' for full-field stimulation of the central retina; 50–90' for hemi-field stimulation of the peripheral retina
    9. Field size: over 8° for full-field stimulation; over 10–16° for hemi-field stimulation
    10. Color: The color of the pattern elements should be specified.
C. Recording parameters
    1. Number of channels: 3–4
    2. Electrode placements:
      a. Midoccipital (MO), 5 cm above the inion
      b. Left and right occipital (LO, RO), 5 cm lateral to MO
      c. Left and right posterior temporal (LT, RT), 10 cm lateral to MO
      d. Midfrontal (MF), 12 cm behind the nasion
      e. Ground electrode at the vertex or elsewhere on the head.
    3. Montages
      a. Full-field stimulation:
        (1) Channel 1: RO–MF
        (2) Channel 2: MO–MF
        (3) Channel 3: LO–MF
        (4) A fourth channel may be used to record the pattern ERG between an infraorbital and a lateral orbital electrode.
      b. Hemi-field stimulation
        (1) Right hemi-field stimulation:
          Channel 1: RO–MF
          Channel 2: MO–MF
          Channel 3: LO–MF
          Channel 4: LT–MF
        (2) Left hemi-field stimulation:
          Channel 1: RT–MF
          Channel 2: RO–MF
          Channel 3: MO–MF
          Channel 4: LO–MF
    4. Filter settings: Low-frequency filters = 0.2–1 Hz; high-frequency filters = 200–300 Hz.
    5. Number of responses averaged: 100–200 for full-field stimulation; 200 for hemi-field stimulation
    6. Sweep length: 250 msec; up to 500 msec in infants and for abnormally delayed VEPs
D. Analysis
    1. Normal peaks
      a. Full-field VEPs: N75, P100, N145 in midline and lateral VEPs
      b. Hemi-field VEPs: N75, P100, and N145 over midline and ipsilateral to the stimulated hemi-field; P75, N105, and P135 contralateral to the stimulated hemi-field
    2. Criteria for abnormal VEPs
      a. Full-field VEPs
        (1) Absence of any peaks
        (2) Abnormally long peak latency of P100

TABLE 7.1 *continued*

   (3) Abnormally long $\overline{P100}$ interocular latency difference
   (4) Questionable criteria: Abnormally long peak latency and intraocular latency difference of $\overline{N75}$, abnormally
     large lateral occipital amplitude ratio of full-field VEPs
  b. Hemi-field VEPs
   (1) Absence of any peaks ipsilateral and contralateral to the stimulated hemi-field
   (2) Questionable criteria: Abnormal latency and waveform

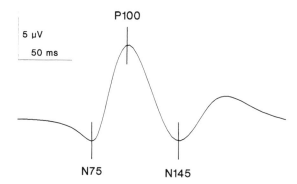

FIGURE 7.1. *Normal midline occipital VEP to monocular full-field pattern reversal stimulation showing $\overline{N75}$, $\overline{P100}$, and $\overline{N145}$ peaks. Stimulation with squares of 30' sidelength in a 14° (horizontal) by 11° (vertical) field of squares reversing at 2/sec. Recording between midoccipital (MO) and midfrontal (MF) electrodes. Positivity at the occiput is plotted upward.*

regardless of whether they are produced by pattern reversal, appearance, disappearance, or flash. The VEPs to checkerboard pattern appearance, disappearance, and flashes and the steady-state VEPs to checkerboard stimuli are presented elsewhere.

## 7.2 SUBJECT VARIABLES

### 7.2.1 Age

#### *7.2.1.1 Before adulthood*

Pattern VEPs can usually not be recorded in young children because they do not fixate well.

However, careful investigations of pattern VEPs starting at preterm age have shown that the latency decreases rapidly during the first year of life and reaches adult values at the end of the first year for check sizes of 50–60' and at over 5 years old for checks of 12–15'. Accommodation in infants is better for larger than for smaller checks.[79] VEP amplitude and latency have been reported to remain stable in older children,[35] although latency and amplitude may decrease during adolescence. Amplitude may continue to increase into the fourth decade.[75]

#### *7.2.1.2 Adults*

The latency of the $\overline{P100}$ increases after the age of 60 years so that age-dependent normal controls are needed to correctly distinguish normal from abnormal in the elderly. Some studies suggest that the $\overline{P100}$ latency increases through the entire adult life.[76] Age effects on latency have been reported to be more prominent for low luminance levels and for small check sizes. Studies on the effect of age on amplitude report a decrease with old age, no change, or an increase.

### 7.2.2 Sex

Women were found to have shorter $\overline{P100}$ latency than men. The difference does not justify separate criteria for abnormality.[1,82]

### 7.2.3 Visual Acuity

A visual acuity of less than 20/200 is likely to reduce VEP amplitude and to increase latency, especially for small check sizes and low contrast.[22,88]

### 7.2.4 Pupillary Size

Normal variations of pupillary size do not affect the VEP to patterned stimuli, but extreme meiosis and mydriasis may alter stimulus luminance and thereby change VEP amplitude and latency.

### 7.2.5 Ocular Dominance

Although the VEP to stimulation of the dominant eye has been reported to have slightly shorter latency and higher amplitude, clinical studies do not use different normative values for dominant and nondominant eye stimulation.

## 7.3 EFFECTS OF STIMULUS CHARACTERISTICS

### 7.3.1 Stimulators

#### 7.3.1.1 Television screens

Television picture tubes under control of pattern-generating circuitry are commonly used. These stimulators may be bought commercially or built from published circuits. TV pattern stimulators have several advantages: They are sufficiently large and bright, can generate checks and horizontal or vertical bars of different sizes and present them at different rates, and may be used to stimulate visual hemi-fields or quadrants without shift of the fixation point. Patterns can be made to reverse or to appear and disappear in alternation with a diffusely illuminated or dark background. The stimulus pattern may be mixed with a TV program, such as a cartoon, to keep young children looking at the screen without significant degradation of the VEP.

The onset of the TV pattern stimulus is slower than that obtained with other pattern generators. This accounts for some latency differences among laboratories, but does not reduce the value of TV stimulation for clinical VEP testing. The slow, gradual onset of the stimulus is due to the way the stimulus image is generated on TV. The change of the TV picture progresses in horizontal lines from the top to the bottom of the screen and is completed in the time required for the presentation of a TV picture frame, namely 16.7 msec at the power line frequency of 60 Hz used in the United States and in 20 msec at the line frequency of 50 Hz used in many other countries. Although the frame rate can be increased, this reduces the sharpness of the image. The gradual appearance of the pattern change on the TV screen leads to a gradual projection of the stimulus pattern on the retina; the commonly used patterns with relatively small check size become effective as a visual stimulus when they reach the central retina. If the pattern change always begins at the top of the screen, the time to its appearance at the fixation point is added to the VEP latency measured from the beginning of the reversal. Although this makes the latency of pattern reversal VEPs longer than that with fast pattern shifts, the difference seems to become insignificant if no fixation point is used, that is, if the interval between pattern change and its appearance at the central retina is allowed to be irregular. However, it is not desirable to lock the pattern change to the onset of the TV frame because this timing also locks the onset of the averaging epoch to the power-line frequency and favors the buildup of power-line artifact in the average. Therefore, the change in pattern is usually unlocked from the start of the TV frame and begins in different parts of the screen with each stimulus. This gives the responses a slightly more variable latency, resulting in a broader VEP peak.

Television pattern generators have minor disadvantages: Contrast borders between dark and light are not absolutely sharp, and the control of contrast and luminance is limited and not linear.

#### 7.3.1.2 Projection with a pivoting mirror

The image of a checkerboard pattern is projected onto a screen by reflecting it off a small mirror mounted on a pen motor or a fast-moving galvanometer. The mirror is rotated intermittently to shift the pattern by exactly the width of one square. This results in a pattern of change that appears in all parts of the retina at the same time and with a speed depending on the duration of the mirror movement. Movements of 5–10 msec give sharply peaked VEPs with latencies of sufficiently small variation for good clinical diagnostic use. This method is quite effective and widely used even though it is less versatile than the TV technique: The choices of square sizes, hemifield and quadrantic presentations, and stimulus rates are limited; contrast is difficult to control.

### 7.3.1.3 Light-emitting diodes

Light-emitting diodes (LEDs) are mounted in a square matrix and connected so that half of them, forming the light squares of a checkerboard pattern, are turned on while the other half, forming the dark squares, are turned off. The electronic switching between the two phases gives the advantage of very fast pattern reversals. Disadvantages are invariant element size, small stimulus field size, colored light, low luminance, and contrast borders that are often not sharp.

### 7.3.1.4 Custom-made stimulators

*7.3.1.4.1 Tachistoscopic displays.* Tachistoscopes can present patterns with switching times of less than 1 msec. They are not in routine use.

*7.3.1.4.2 Two projectors with synchronized shutters.* Slides showing patterns may be projected alternately on a screen from two projectors with electronically controlled shutters having switching times of a few milliseconds. Pattern reversal is produced by projecting complementary checkerboard patterns; pattern appearance and disappearance are generated by alternately projecting a pattern and a uniform gray field.

*7.3.1.4.3 Patterned mirrors.* Mirrors with alternating reflecting and transparent elements have been used in conjunction with independently switchable light sources to project reversing or appearing and disappearing checkerboard patterns.

*7.3.1.4.4 Polarized light.* Pattern reversal stimuli may be generated by passing light through a pattern of Polaroid squares with alternating horizontal and vertical polarizing axes and by viewing the slide through a rotating Polaroid disc; this generates pattern reversals with a sinusoidally changing time course, suitable for steady-state VEPs.

*7.3.1.4.5 Maxwellian systems.* The stimulus may be viewed through an eyepiece that focuses the light from a pattern so that it forms an image directly in the eye. Upon entry into the eye, the light waves are almost perfectly convergent. Therefore, the image is independent of pupillary size. This system differs from others in that the image is directly projected onto the retina.[90]

## 7.3.2 Stimulus Rate and Phase

Transient checkerboard pattern reversal VEPs are usually elicited at 2/sec. Slower rates of stimulation produce no change of the VEP. An increase of stimulus rate to 4/sec may significantly increase the latency of the transient VEP.

The stimulus rate of pattern reversal stimuli equals twice the alternation or stimulus pulse rate because each complete alternation or stimulus pulse generates two stimuli, namely, the transition of the pattern from one phase into the other and the return to the original phase. Responses to transitions into both phases are usually averaged together because these responses are similar.

## 7.3.3 Contrast

### 7.3.3.1 Sharpness of contrast borders

Blurring of contrast borders degrades the pattern VEP.[80] Reduction of amplitude and increase of latency are especially prominent with small check sizes. Because blurring may be caused by refractive errors, these should be corrected so that they will not interfere with the evaluation of conduction through the central visual pathway. As a rule, a decrease of visual acuity does not alter VEPs to checkerboard pattern stimulation unless it reaches 20/200 or unless the smallest normally effective check sizes are used.

### 7.3.3.2 Effect of contrast depth

A reduction of the contrast difference between the light and dark squares, usually set to 0.5 or higher, may increase the latency and reduce the amplitude of the VEP, but the range and the degree of this effect depend on many other variables including luminance and size of the squares. An increase of contrast above the fairly high levels usually used in clinical VEP work does not change the VEP.

## 7.3.4 Luminance

An increase of the mean luminance of a checkerboard pattern stimulus decreases the latency and increases the amplitude of the VEP to a degree that depends on other variables, including check size and contrast. Under commonly used conditions, decreasing the mean luminance of a reversing checkerboard pattern by a factor of 10 increases its amplitude by about 15%. The luminance of light and dark pattern elements used in different laboratories varies widely: The light squares used in some studies are darker than the dark squares in others.

## 7.3.5 Check Size and Field Size

A decrease of check size increases VEP amplitude to a maximum at a visual angle of 10–15′ and increases latency, especially at the smallest effective check sizes. The effect of check size varies in different parts of the retina. The central 4–5° of the retina, or fovea, generates the largest part of the VEP. The fovea is most sensitive to small check sizes of about 10′ whereas the peripheral parts of the retina are more sensitive to larger check sizes; locations up to 7.5° from the fovea are susceptible to progressively increasing sizes. The check size more effective at a given eccentricity depends on the stimulus field size. Check sizes exceeding those that are effective as pattern stimuli in a particular part of the retina produce responses mainly due to local retinal luminance changes. The crossover between pattern and luminance effects depends on many variables, such as location of the stimulus on the retina, brightness contrast, orientation of the checks, luminance, and level of light adaptation. Because of the different effect of small and large check sizes on different parts of the retina, small checks need to be presented only in a small central field, whereas large checks require larger fields to cause their pattern effect in the peripheral retina; while they act as pattern stimuli at their respective effective sites, small checks have little effect in the peripheral retina, but large checks cause luminance in the central retina. The smallest effective check size of about 10′, although producing the largest VEP under ideal conditions, is not used in routine clinical studies because blurring, often caused by refractive errors, affects latency and amplitude of the VEP to the smallest size much more than those of VEPs to larger sizes.

## 7.3.6 Location of the Stimulus in the Visual Field

### 7.3.6.1 VEPs to full-field and hemi-field stimulation

Stimulation of the left and right hemi-fields of the retina produces VEPs different from those by full-field stimulation. This difference is small if only the central part of the retina is stimulated with checks of small size presented in a small field. It becomes larger if the periphery of the retina is stimulated with checks of 50–90′ in a field of over 16°. The difference between full-field and hemi-field VEPs is greatest in recordings that include lateral occipital or posterior temporal electrode placements.

### 7.3.6.2 VEPs to upper and lower retinal stimulation

Stimulation of the upper hemiretina seems to produce VEPs of shorter latency and more anterior distribution than does stimulation of the lower hemiretina. Clinical studies usually do not use this variation but place the fixation point in the middle of the vertical extent of the stimulus pattern.

## 7.3.7 Monocular versus Binocular Stimulation

VEPs to stimulation of either eye are normally very similar to each other. The VEP to stimulation of both eyes may have slightly larger amplitude but usually has the same latency as VEPs to stimulation of each eye. In patients with an abnormal VEP to stimulation of one eye, the VEP to binocular stimulation is usually normal.

### 7.3.8 Effect of Pattern Orientation

Because the major spatial Fourier component of checkerboard patterns is oriented diagonally to the checks, checkerboards are more effective when presented as diamonds than as squares.[55] However, squares are used in clinical testing because they are easier to generate.

---

## 7.4 EFFECT OF RECORDING PARAMETERS

### 7.4.1 Electrode Placements and Combinations for Full-Field and Hemi-Field VEPs

#### 7.4.1.1 VEPs to full-field stimulation

Normal full-field pattern VEPs have a maximum at the midline of the head and are usually fairly symmetrical on the two sides, but may be much larger on one side than the other. The lateral extension of the VEP varies considerably

in normal subjects: The amplitude of the $\overline{P100}$ drops off to the sides much more rapidly in some subjects than in others. The amplitude of VEP recordings further depends on the choice of the reference electrode and may be low with ear or vertex references because similar potential changes may appear both at the occipital and the reference electrode and lead to partial cancellation of the VEP due to differential amplification. Some normal subjects show no VEPs in recordings between occipital and vertex electrodes because of complete cancellation. For full-field studies of pattern VEPs, recordings are made in three or four channels. To conform with placements commonly used for hemi-field studies, recording electrodes are placed on a coronal line 5 cm to the left and right. A midfrontal electrode is positioned 12 cm above the bridge of the nose. A ground electrode is placed somewhere on the head, such as the vertex. Four-channel recordings should use three channels to record between each of the three occipital electrodes and the midfrontal electrode. The fourth channel may be used to record the VEP between the midoccipital and a vertex electrode or to record the ERG.

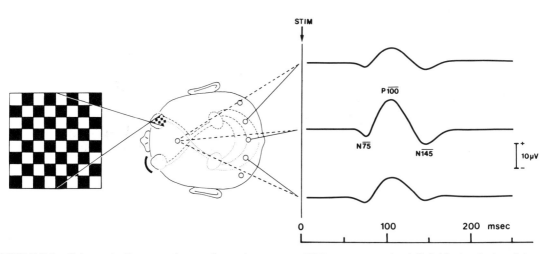

FIGURE 7.2. *Schematic diagram of normal transient pattern VEPs to monocular full-field stimulation. Stimulation of one eye produces VEPs that are distributed approximately symmetrically over both occipital areas with a maximum at the midline. They have a major positive peak (P100), preceded and followed by smaller negative peaks (N45, N145). Positivity at the occiput is plotted upward.*

### 7.4.1.2 VEPs to hemi-field stimulation

Normal hemi-field VEPs to checkerboard stimulation have a paradoxical distribution that is best seen with a technique that uses (1) stimulation with fairly large check sizes of about 1°, presented in a field sufficiently large to affect the retinal periphery, and (2) recordings from midline and lateral occipital, or posterior temporal, electrodes in reference to a midfrontal electrode. With this technique, stimulation of one visual hemi-field, which excites the opposite hemisphere, produces a VEP that is largest in scalp recordings from the midline and the side of the stimulated hemi-field; these recordings show N$\overline{75}$, P$\overline{100}$, and N$\overline{145}$ peaks resembling those of the full-field VEP (Figures 7.3 and 7.4). On the opposite side, that is, over the excited hemisphere, the VEP usually has lower amplitude (Figure 7.3*b–f*). Recordings from lateral occipital or posterior temporal electrode positions may show a VEP with a shape resembling a mirror image of the contralateral VEP, having P$\overline{75}$, N$\overline{105}$, and P$\overline{135}$ peaks (Figure 7.4). Some subjects have symmetrical VEPs to hemi-field stimulation. Like full-field VEPs, hemi-field VEPs extend laterally to a degree that varies among different subjects.

The full-field VEP consists of the sum of the two asymmetrical hemi-field VEPs (Figure 7.5).

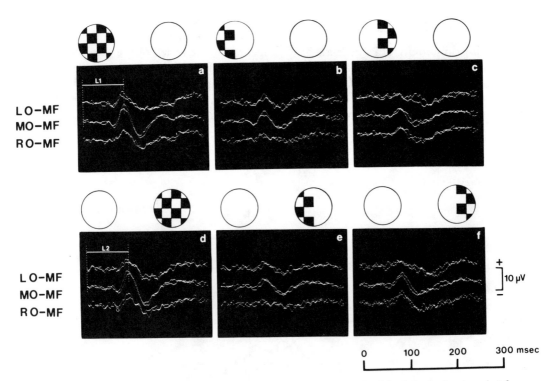

FIGURE 7.3.   *Normal midline and lateral occipital VEPs to stimulation of the left eye (top) and right eye (bottom) with full-field (a, d), left hemi-field (b, e), and right hemi-field (c, f) checkerboard pattern reversal. The latencies measured to P$\overline{100}$ of the midline VEP (L1, L2) are normal. The amplitudes of the half-field VEPs ipsilateral to the stimulated half-field are higher than those on the other side. A P$\overline{105}$ is not seen in this case in which posterior temporal electrodes were not used. Stimulation with checkerboard patterns of squares of 30′ (a, d) and 60′ (b, c, e, f) sidelength. Recordings between left occipital (LO), midoccipital (MO), right occipital (RO), and midfrontal (MF) electrodes. Positivity is plotted upward.*

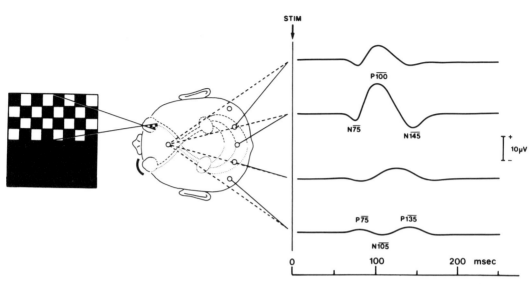

FIGURE 7.4. *Schematic diagram of normal transient pattern VEPs to monocular hemi-field stimulation. Stimulation of each hemi-field produces occipital VEPs that have maximum amplitude, and peaks similar to those of full-field VEP, at the midline and ipsilateral to the stimulated hemi-field, that is, opposite the excited hemisphere; VEPs opposite the stimulated hemi-field, that is, over the excited hemisphere, usually have lower amplitude and may show P75, N105, and P135 peaks. Because most averagers have no more than four channels and can not record from all five posterior electrodes simultaneously, routine studies may exclude recordings from the posterior temporal electrode ipsilateral to the stimulated hemi-field. Positivity at the occiput is plotted upward.*

The predominance of the hemi-field VEP on the side of the stimulated field is thought to be due to the spatial orientation of the visual cortex in the depth of the occipital lobe: Excitation of the visual cortex on one side seems to generate an electric field of higher amplitude on the other side of the head.[4] However, this paradoxical distribution is not seen with all stimulating and recording methods. VEPs with a maximum over the excited hemisphere may be recorded with different recording electrode placements and with stimulation using small field sizes, pattern onset rather than reversal, diffuse light, or repetition rates producing steady-state VEPs.

Most hemi-field studies now use four recording channels. Three channels are connected to the same electrode combinations as used for full-field studies. An additional left posterior temporal electrode (LT) and right posterior temporal electrode (RT) are placed 10 cm lateral to the midline occipital electrode, on the coronal line of occipital electrodes 5 cm above the inion. The fourth channel is used to record between the posterior temporal electrode contralateral to the stimulated hemi-field and the midfrontal electrode (Table 7.1, C.3). If more than four channels are available, recordings may be made between all five posterior electrodes and the midfrontal electrode. Other electrode placements, such as those of the International 10–20 System, are used occasionally. Two-channel recordings are barely sufficient for hemi-field studies. Bipolar montages of electrode chains across the occipital areas produce distortions of VEPs.

### 7.4.2 Other Recording Parameters

Most other recording parameters are indicated in the general section (6.4). The number of recording channels should be at least three for full-field stimulation and at least four for hemi-field stimulation. The number of responses in

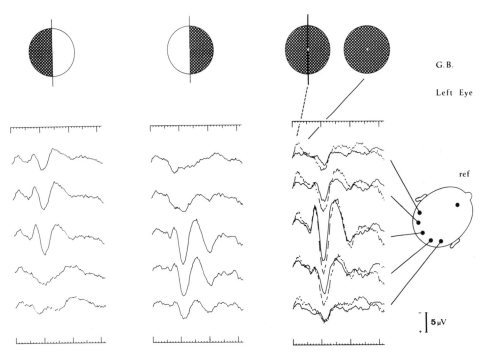

**FIGURE 7.5.** *Distribution of transient pattern shift VEPs to hemi-field and full-field stimulation in midline and lateral occipital recordings with a frontal reference electrode. Stimulation of the left hemi-field (left column) and right hemi-field (middle column) of the left eye with alternating checkerboard patterns of large check and field sizes produces asymmetrical VEPs with a maximum in the midline and ipsilateral to the stimulated hemi-field. Full-field stimulation (right column, solid lines) produces fairly symmetrical VEPs with a maximum at the midline; these full-field VEPs resemble the algebraic summation of the asymmetric hemi-field VEPs (right column, dashed lines). Electrode placements as indicated in the head diagram; midfrontal reference electrode. Negativity at the occiput is plotted upward; the P100 points down. From Blumhardt et al. (Br. Ophthalmol 61:454, 1977) with permission of the authors and British Medical Journal.*

each average is about 100 for full-field VEPs and about 200 for hemi-field VEPs.

## 7.5 PRECHIASMAL, CHIASMAL, RETROCHIASMAL, AND OPHTHALMOLOGICAL STRATEGIES

Different combinations of monocular and hemi-field stimulation and of recording from midline and lateral occipital electrodes are used to detect lesions in the three major portions of the visual pathway, namely (1) the prechiasmal part, consisting of retina and optic nerve, (2) the optic chiasm, and (3) the retro-chiasmal portion, consisting of optic tract, lateral geniculate body, optic radiation, and visual cortex.

### 7.5.1 Prechiasmal Strategy

To test the prechiasmal parts of the visual pathway, one must stimulate each eye separately and record from the midline of the occiput; lateral occipital recordings are often added (Figure 7.6). Because prechiasmal studies aim at large full-field VEPs to stimulation of the central retina, they require fields of no more than about 10° in combination with centrally more effective check sizes of about 30′.

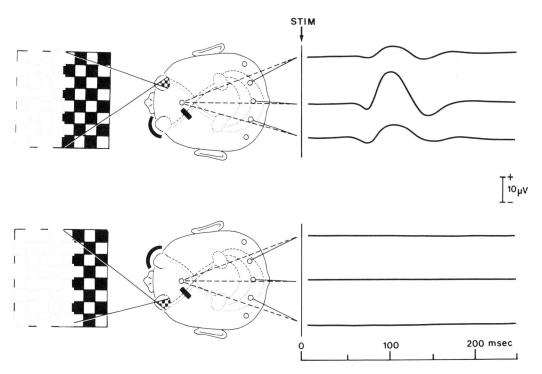

FIGURE 7.6. *Strategy for detecting prechiasmal conduction defects by monocular full-field stimulation and midline recording. Top: Stimulation of the right eye and occlusion of the left eye produces a normal VEP. Bottom: Stimulation of the left eye and occlusion of the right eye produces no VEP because of a lesion completely interrupting the left optic nerve. Incomplete lesions may increase the latency and reduce the amplitude. Positivity at the occiput is indicated by an upward deflection.*

With this strategy, a prechiasmal lesion on one side is detected by an abnormality of the VEP to stimulation of the eye on that side (Figure 7.6). Whereas complete interruption of an optic nerve abolishes the monocular VEP, partial lesions usually increase the latency and decrease the amplitude of the VEP. A prechiasmal lesion affecting only the nasal or temporal half of the retina or optic nerve of one eye may cause an abnormal VEP only on stimulation of the opposite hemi-field of that eye.

Lesions of both optic nerves or eyes may produce VEP abnormalities to stimulation of each eye. Such bilateral monocular abnormalities often differ from each other. If they are similar, they must be distinguished from those produced by chiasmal and retrochiasmal lesions; testing with hemi-field stimulation is needed in these cases. A reduction of VEP ampli-

tude is not always due to lesions: Symmetrically low amplitude, or even a slight difference of amplitude between monocular midline VEPs to stimulation of each eye, may be seen in some normal subjects and can therefore not be used as a reliable indicator of a prechiasmal lesion.

The prechiasmal strategy yields abnormal VEPs with lesions both of the eye and of the optic nerve. Although VEPs can not always distinguish between these two types of lesions, ocular lesions generally reduce VEP amplitude without affecting latency, but marked increases of latency are caused only by optic nerve lesions, especially by demyelination. However, a slight increase of latency may have ocular causes, whereas some optic nerve lesions, such as tumors and ischemia, may be manifested mainly by a decrease of VEP amplitude; complete absence of an EP may be due to either ocular or

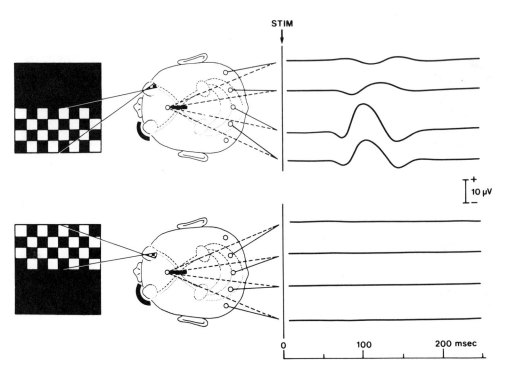

FIGURE 7.7. *Strategy for detecting chiasmal lesions by monocular hemi-field stimulation and bilateral occipital recording. Top: Stimulation of the nasal hemi-field of the right eye, exciting the temporal half of the eye and right hemisphere, produces normal VEPs over both hemispheres, larger on the left side. Bottom: Stimulation of the temporal hemi-field of the right eye, exciting the nasal half of the eye, produces no VEPs on either side because of a lesion completely interrupting the optic nerve fibers that cross in the chiasm. Study of the other eye would show no VEPs to temporal hemi-field stimulation and normal VEPs to nasal hemi-field stimulation. Incomplete lesions may reduce the amplitude, distort the shape, and occasionally increase the latency of the involved VEPs.*

nerve lesions. Several other characteristics may help to distinguish ocular from optic nerve lesions.

### 7.5.2 Chiasmal and Retrochiasmal Strategies

To detect a lesion at or behind the chiasm, one must stimulate the right and left visual hemi-fields separately and record VEPs from the occipital areas of each hemisphere. Hemi-field stimulation is necessary because, contrary to expectations, a lesion of the chiasm or of one side of the retrochiasmal pathway may produce inconclusive or normal VEPs even in recordings from each side as long as full-field stimuli are used. The reason for this is probably that the normal VEP generated by the intact hemisphere, distributed rather widely over both sides of the head in most persons, may overshadow the unilateral abnormality; furthermore, normal full-field VEPs may be fairly asymmetrical. Therefore, to detect more reliably chiasmal and retrochiasmal abnormalities, each hemisphere must be excited separately by stimulating either the right or left hemi-field while recording from each side of the occiput. Because retrochiasmal studies attempt to excite the peripheral halves of the retina separately, they require larger stimulus fields and use the peripherally more effective larger check sizes of 50–90′.

When studied with this method, chiasmal lesions, which characteristically interrupt fibers from the nasal half of each eye, cause abnormalities of the VEPs to stimulation of the temporal hemi-fields of each eye. Fibers from the nasal hemi-fields remain intact and produce VEPs which, although generated on the side of the stimulated eye, have a maximum over the hemisphere opposite that eye (Figure 7.7). Retrochiasmal lesions affect the VEP to stimulation of the contralateral visual hemi-field of each eye (Figure 7.8). Thus, a left-sided retrochiasmal lesion may abolish the VEPs to stimulation of the right visual hemi-field, which is normally represented maximally over the right, uninvolved side; this lesion leaves intact the VEP to stimulation of the left visual hemi-field, which has a maximum over the left, involved hemisphere.

Hemi-field stimulation can detect only massive lesions that completely abolish the VEP to stimulation of the opposite hemi-fields; such lesions cause large field defects, usually fairly complete hemianopias. Less extensive lesions, for instance those causing only quadrantic field defects, are not detected reliably, and lesions of the posterior hemispheres not causing any field defects do not produce VEP abnormalities.[10]

Full-field stimulation, in combination with lateral occipital recordings, may reveal VEP

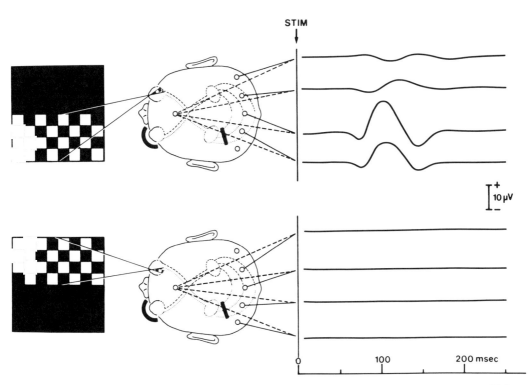

FIGURE 7.8. *Strategy for detecting retrochiasmal lesions by monocular hemi-field stimulation and bilateral occipital recording. Top: Stimulation of the nasal half-field of the right eye, exciting the temporal half of the eye and the right hemisphere, produces normal VEPs over both hemispheres, larger on the left side. Bottom: Stimulation of the temporal hemi-field of the right eye, exciting the nasal half of the eye, produces no VEPs on either side because of a lesion completely interrupting the fibers of the left optic radiation. Study of the other eye would show normal VEPs to temporal hemi-field stimulation and no VEPs to nasal hemi-field stimulation. Incomplete lesions may reduce the amplitude, distort the shape, and occasionally increase the latency of the involved VEPs.*

asymmetries similar to those obtained by stimulation of the intact-preserved hemi-field. Retrochiasmal lesions leave intact VEPs with a maximum on the side of the lesion and therefore cause an abnormality that does not change sides with the stimulated eye, that is, an uncrossed asymmetry. Chiasmal lesions leave intact VEPs opposite the stimulated eye and thereby produce a crossed asymmetry.[11]

Although chiasmal and retrochiasmal lesions mainly alter VEP amplitude, they occasionally produce an increase of latency that may be recorded with the prechiasmal strategy and could be confused with prechiasmal lesions if not evaluated with the retrochiasmal strategy.

### 7.5.3 Choice of Strategies

Ideally, all three portions of the central visual pathway should be examined if an abnormality is suspected in any one of them. Clinically most useful is the prechiasmal strategy: It can detect lesions not found with other methods of examination; the chiasmal and retrochiasmal strategies are much less sensitive, giving abnormal findings usually only in cases that also show large visual field defects. Although full-field studies are most rewarding and usually provide sufficient evidence for a prechiasmal lesion if the VEP abnormality is clearly monocular, lesions of the chiasmal and retrochiasmal portions must be considered in many cases of abnormal full-field VEPs, especially those showing bilateral monocular abnormalities; these cases require further investigation with hemi-field studies. Furthermore, hemi-field stimulation may increase the sensitivity of prechiasmal studies: Abnormal VEPs may be found on stimulation of only one or the other hemi-field of one eye in cases with normal full-field VEPs and may suggest prechiasmal lesions restricted to the nasal or temporal side of the optic nerve or eye. Some laboratories therefore use both full-field and hemi-field studies in every patient.

### 7.5.4 Ophthalmological Strategies

VEPs are sometimes used to study ocular lesions, especially those producing visual defects. VEPs have been used to answer the question whether a person can see. These studies are based on the general rule that VEPs are present in subjects who can see and absent in subjects who can not see. However, VEPs may be present and vision absent in patients with cortical blindness, whereas VEPs may be absent and vision present in several conditions including optic neuropathies, equipment failure, and unintentional or intentional lack of focusing on a patterned stimulus. Therefore VEPs can not be reliably used to determine whether a patient can see. Nevertheless, the presence of a normal VEP to pattern stimuli in the absence of bilateral occipital lesions strongly suggests that a subject is able to see, however, the absence of VEPs, characteristic of all kinds of blindness except cortical blindness, is by no means diagnostic of blindness.

VEPs have also been used to study ocular lesions, visual acuity, and refraction, especially in infants and other noncommunicating subjects. Visual acuity can be estimated by determining the size of pattern elements giving VEPs of maximum amplitude or by determining the smallest size of pattern elements capable of producing a VEP; refractive errors and astigmatism have been estimated by selecting corrective lenses that give the largest VEP. Other ophthalmological applications include studies of brightness contrast sensitivity, color vision, adaptation, and binocularity.

# 8

# The Abnormal Transient VEP to Checkerboard Pattern Reversal

## 8.1 CRITERIA FOR ABNORMAL PATTERN REVERSAL VEPS

### 8.1.1 Absence of VEPs

The complete absence of any peak larger than residual noise is abnormal and indicates a clinical defect if lack of focusing and technical problems are excluded. To demonstrate the absence of even a very delayed VEP, an analysis time of 500 msec should be used if no peak appears in recordings with shorter analysis times.

### 8.1.2 Abnormally Long $P\overline{100}$ Latency

The latency of the $P\overline{100}$ is measured between the pattern reversal, usually coinciding with the beginning of the analysis period, and the peak of the first major occipital positive peak.

### 8.1.3 Abnormally Long Interocular $P\overline{100}$ Latency Difference

The interocular latency difference is determined by subtracting the $P\overline{100}$ latency of the midline VEP to stimulation of one eye from the $P\overline{100}$

latency of the midline VEP to stimulation of the other eye. This value may exceed the limit of normal interocular latency differences even if both absolute latencies remain within the normal range.

### 8.1.4 Abnormally Long $N\overline{75}$ Latency

The latency of the $N\overline{75}$ is often increased in cases also showing a delayed $P\overline{100}$, but an abnormal delay of the negative peak alone can merely raise a suspicion of abnormal conduction.

### 8.1.5 Abnormally Large $P\overline{100}$ Amplitude Ratio

A lateral occipital amplitude ratio may be calculated by measuring $P\overline{100}$ amplitudes of VEPs on the two sides of the head and dividing the larger $P\overline{100}$ amplitude by the smaller one. The normal amplitude ratio for full-field VEPs varies considerably in and among subjects. Amplitude ratios for half-field VEPs indicate higher amplitude over the hemisphere on the side of the stimulated hemi-field in most subjects.

## 8.2 GENERAL CLINICAL INTERPRETATION OF ABNORMAL VEPS TO CHECKERBOARD PATTERN REVERSAL

Abnormal VEP findings must be interpreted with regard to the possible location and type of the underlying defect. Before VEP abnormalities can be accepted as indicators of lesions in the visual pathway, several technical problems must be excluded. Lesions affecting the prechiasmal, chiasmal, and retrochiasmal parts of the visual pathway can be recognized by fairly characteristic constellations of VEP abnormalities. Abnormal VEP findings and their clinical implications for full-field stimulation are outlined in Table 8.1, and for hemi-field stimulation in Table 8.2.

### 8.2.1 Technical and Ocular Problems and Lack of Focusing

Absence of VEPs may be due to various technical problems. Such problems must be suspected especially in cases where no VEP can be elicited by full-field and half-field stimulation of either eye with check sizes of 1°, even though the subject can see the stimulus pattern. Technical

problems are less likely to explain the absence of a monocular full-field or half-field VEP if stimulation of the other eye or half-field produces a VEP. Common technical problems are failure to synchronize the stimulator with the sweep of the averager, and faulty recording electrodes, amplifiers, and averaging channels.

VEP abnormalities may be the result of various extraocular and ocular lesion: Ptosis, cataracts, and extreme meiosis may reduce illu-

TABLE 8.3.  Clinical interpretation of hemi-field VEP abnormalities

| Abnormality | Interpretation |
| --- | --- |
| Abnormal VEP to stimulation of both temporal hemi-fields | Chiasmal lesion |
| Abnormal VEP to stimulation of both left or right hemi-fields | Retrochiasmal lesion |
| Abnormal VEP to stimulation of hemi-field of one eye | Optic nerve or eye lesion; rule out chiasmal or retrochiasmal lesion |

Table 8.2.  Clinical interpretation of monocular full-field VEP abnormalities

| Worst eye | Best eye | Interpretation |
| --- | --- | --- |
| Absent VEP | Absent VEP | Technical problems; bilateral prechiasmal, chiasmal, or retrochiasmal lesions |
| Absent VEP | Increased latency | Bilateral optic nerve lesions, chiasmal lesion |
| Absent VEP | Very low amplitude VEP | Suspect bilateral prechiasmal or chiasmal lesion |
| Absent VEP | Normal | Optic nerve or ocular lesion |
| Increased latency | Low amplitude | Suspect bilateral prechiasmal or chiasmal lesion |
| Increased latency | Normal | Optic nerve lesion |
| Normal latency but increased interocular difference | Normal | Optic nerve lesion |
| Very low amplitude | Low amplitude or increased latency | Suspect bilateral prechiasmal, chiasmal, or retrochiasmal lesions; may be normal |

mination of the retina; corneal opacities, cataracts, retinal lesions, and refractive errors may interfere with the sharp projection of small pattern elements onto the retina. Although reduced luminance and blurring of a patterned stimulus usually reduce VEP amplitude, they may also slightly increase VEP latency. However, in contrast to optic nerve lesions, ocular lesions produce no more than a slight increase in latency. As a rule, ocular defects can be distinguished from optic nerve lesions in that they alter the VEP only if they also interfere with vision, that is, if they reduce visual acuity or cause central scotomata; the VEP may remain normal with scotomata that spare the central 3°. Ocular lesions causing abnormal VEPs can usually be detected by ophthalmologic examination. A few ophthalmologic disorders have been the subject of VEP studies. ERGs to diffuse and patterned light may be of help.

Another possible cause of reduced VEP amplitude or slightly prolonged latency is unintentional or intentional lack of focusing on the stimulus pattern.

## 8.2.2 Abnormal VEPs to Monocular Full-Field Stimulation

### 8.2.2.1 Increase of latency

A significant increase of $\overline{P100}$ latency to stimulation of one eye or an increased $\overline{P100}$ latency difference between VEPs to stimulation of each eye practically always indicates an optic nerve lesion on the side of the longer latency, especially if stimulation of the other eye produces a normal VEP. A prolonged latency of VEPs to stimulation of either eye usually indicates lesions of both optic nerves; rarely is it caused by chiasmal or retrochiasmal lesions. The magnitude of the latency increase may differ for both eyes in prechiasmal and chiasmal lesions, but it is similar for both eyes in retrochiasmal lesions. Bilateral optic nerve lesions caused by degenerative and metabolic disorders cause bilateral delays that often, but not always, have the same magnitude on the two sides. Demyelinating lesions are more likely than other lesions to produce significant unilateral or bilateral increases of VEP latency.

### 8.2.2.2 Absence or reduced amplitude of VEPs

Absence of a VEP to monocular stimulation is usually abnormal and due to an ipsilateral prechiasmal lesion involving either the eye or the optic nerve; if the VEP to stimulation of the other eye is also abnormal, it is necessary to consider bilateral prechiasmal lesions, a chiasmal lesion, and bilateral retrochiasmal lesions. The possibility of technical problems has to be carefully eliminated, especially in cases of bilateral absence of VEPs. Very low amplitude of a VEP is a less reliable indicator of lesions than is increased latency, but may occur in ocular lesions and optic nerve compression, ischemia, or injury; low amplitude may be combined with increased latency. A very marked reduction of a VEP recorded from a lateral occipital electrode, causing an abnormal lateral occipital amplitude ratio, suggests the possibility of a retrochiasmal lesion.

## 8.2.3 Abnormal VEPs to Half-Field Stimulation

### 8.2.3.1 Abnormal VEPs to stimulation of the temporal half-fields

Absence of VEPs to stimulation of the temporal half-fields suggests a chiasmal lesion and is usually associated with bitemporal visual field defects. Reduced amplitude, increased latency, and abnormal distribution of these VEPs are less reliable indicators of clinical lesions.

### 8.2.3.2 Abnormal VEPs to stimulation of corresponding half-fields

Absence of VEPs to stimulation of both left or right half-fields strongly suggests a retrochiasmal lesion opposite the stimulated half-field and is usually associated with homonymous field defects. Reduced amplitude, increased latency, and abnormal distribution of these VEPs suggest the possibility of a retrochiasmal lesion.

### 8.2.3.3 Abnormal VEPs to stimulation of a monocular half-field

Absence of or increased latency of a VEP to stimulation of a half-field of only one eye may suggest a partial lesion of the eye or optic nerve on the side opposite the stimulated half-field.

### 8.2.4 Ophthalmological Problems

Reduced amplitude with or without a slight increase of latency may be seen in ocular disorders such as corneal opacities, cataracts, and glaucoma. Stimulation with patterns of different check size may reveal an increase in the check size causing the largest VEP; this is usually associated with reduced visual acuity. In patients with poor visual acuity due to refractive errors, the amplitude of VEPs to smaller pattern elements may be raised with corrective lenses and give an estimate of the magnitude of the refractive error.

## 8.3 NEUROLOGICAL DISORDERS THAT CAUSE ABNORMAL PATTERN REVERSAL VEPS

Most disorders can be divided easily into prechiasmal, chiasmal, and retrochiasmal. Some disorders, such as multiple sclerosis (MS), leukodystrophies, and renal encephalopathies, may involve more than one segment of the central visual pathway and are here arbitrarily classified according to the site of their lesions most often studied with VEPs. Table 8.3 presents a summary of disorders commonly studied with VEPs and the expected abnormalities.

### 8.3.1 Prechiasmal Lesions

VEPs can be very useful in the diagnosis of MS because they can detect clinically silent conduction defects in the optic nerve and thereby indicate the presence of multiple lesions. VEPs may also help in the diagnosis of tumors compressing the optic nerve and in the diagnosis of ischemic optic neuropathies. VEPs are occasionally used in patients with other diseases that sometimes involve the optic nerve; in these instances, VEPs are usually used not to make the diagnosis of the disease but to determine whether the optic nerve is involved by it. In hereditary conditions involving the optic nerve, an abnormal VEP may be the only or the earliest evidence that a person at risk is affected. In some of the diffuse disorders discussed here, the distinction between prechiasmal and chiasmal or retrochiasmal involvement is difficult to make; chiasmal and retrochiasmal lesions rather than prechiasmal lesions may be partly or entirely responsible for VEP abnormalities, especially those appearing on stimulation of either eye.

#### 8.3.1.1 Retrobulbar neuritis

Acute retrobulbar neuritis, usually occurring in patients with MS, markedly reduces or abolishes VEPs to stimulation of the eye on the affected side (Figure 8.1).[40] On recovery, the amplitude increases as visual acuity improves and eventually may become normal. However, the latency remains abnormally prolonged in most cases of retrobulbar neuritis and MS, although it may return to the normal range in some subjects.

#### 8.3.1.2 Multiple sclerosis

Transient VEPs to alternating checkerboard stimulation are abnormal in a high percentage of patients with MS. The rate of abnormality is high in patients who have a history of retrobulbar neuritis or findings of optic pallor, reduced visual acuity, and visual field defects, but abnormal VEPs are also found in patients without any history or clinical findings suggesting involvement of the visual system. Even though abnormal VEPs are most common in patients with definite MS, less common in probable MS, and least common in possible MS, they are clinically most useful when the diagnosis is in doubt. The VEP may help to distinguish the spinal form of MS from transverse myelitis of other causes. Follow-up studies have confirmed the diagnostic value of EPs.

The most common and diagnostically most important VEP abnormality in MS is an increase of $\overline{P100}$ latency (Figure 8.2), which in many cases is much longer than the delays found in other diseases. This delay may be associated with decreased amplitude and increased duration of the $\overline{P100}$ (Figure 8.2), suggesting dispersion of impulse traveling in abnormal nerve fibers. Complete absence of a VEP is rare except

TABLE 8.4. VEPs in various disorders

| Disorder | VEP findings |
|---|---|
| AIDS | Usually normal |
| Alcoholism | Increased latency in some |
| Charcot-Marie-Tooth disease | Delayed in most patients |
| Chronic renal failure | Delayed VEP, especially in patients on hemodialysis |
| Corneal opacity, miosis, cataract | Reduced amplitude and may be slight increase in latency |
| Diabetes | May be delayed even in clinically asymptomatic patients |
| Down's syndrome | Delayed and low amplitude |
| Endocrine orbitopathy | Delayed VEP |
| Friedreich's ataxia | Increase $\overline{P100}$ latency; low amplitude |
| Glaucoma | Latency may be slightly increased |
| Hereditary ataxia | Conflicting data; usually normal |
| Hereditary spastic paraplegia | Latency may be increased or normal |
| Hysterical blindness | Normal |
| Ischemic optic neuropathy | Reduced amplitude; some latency prolongation though not marked |
| Leber's optic neuropathy | Increased latency, decreased amplitude; VEP is lost eventually |
| Leukodystrophies | Abolish or delay VEP |
| Mitochondrial myopathy | Delayed; may be subclinical |
| Multiple sclerosis | *See* Retrobulbar neuritis. May show abnormalities in absence of clinical symptoms |
| Neurosyphilis | Delayed in 20% of patients |
| Optic nerve tumors | Reduced amplitude or absent; increased latency, though not as marked as with MS |
| Parkinson's disease | Delayed $\overline{P100}$ mainly in patients with associated dementia |
| Pernicious anemia | Slightly delayed in a few cases |
| Phenylketonuria | Increased latency in a few |
| Retinopathy | May have increased latency and decreased amplitude |
| Retrobulbar neuritis | Increased latency and reduced amplitude from affected sides |
| Sarcoidosis | May have increased latency even with no evidence of brain or eye involvement |
| Trauma to optic nerve | Reduced amplitude |

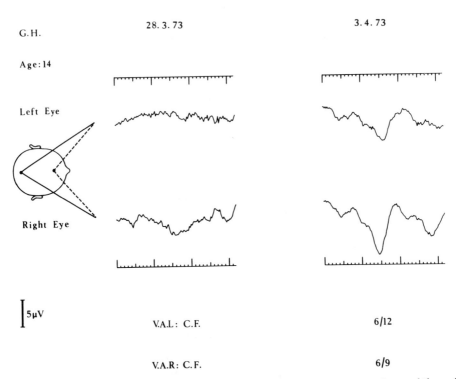

**FIGURE 8.1.** *Transient pattern shift VEPs recorded 3 and 10 days after the onset of acute bilateral optic neuritis in a 14-year-old girl. The first recording (left) shows no peaks to stimulation of the left eye and a questionable peak to stimulation of the right eye; visual acuity of the left and right eye (V.A.L., V.A.R.) was reduced to finger counting. In the second recording (right), both VEPs had returned but were abnormally delayed. Visual acuity had greatly improved. Stimulation of the left eye (upper tracings) and right eye (lower tracings). Recording between midoccipital and midfrontal electrodes. Negativity at the occiput is plotted upward; the P100 points down. Time scale 10, 50, and 100 msec. From Halliday and McDonald (Br Med Bull 33:21, 1977) with permission by the authors and Churchill Livingstone Medical Division of Longman Group Ltd.*

in patients with acute retrobulbar neuritis or severely impaired vision. VEP abnormalities are often monocular but may be binocular. If binocular, they usually are different on the two sides but sometimes are similar or identical (Figure 8.3). Although binocular abnormalities are most commonly due to binocular optic nerve lesions of MS, they may also be due to chiasmal or retrochiasmal lesions, which should be excluded with partial-field stimulation. The use of stimuli of lower luminance has been suggested to increase the detection rate of abnormal VEPs in MS, but this is controversial and not in routine use. Similarly, testing with checkerboard

patterns of different orientation has been reported to increase the incidence of abnormal findings.

The degree of the prolongation of latency may vary over time corresponding somewhat with visual function. As in patients having retrobulbar neuritis without evidence of MS, latency may normalize after months or years in some cases of MS. The amplitude of the VEP may transiently decrease in patients experiencing deterioration of vision as a result of physical exercise. An increase of body temperature alone may also reduce the VEP amplitude in some patients to a greater degree than in normal subjects; heating increases the incidence of abnormal VEP findings in patients with MS.[67]

The diagnostic value of checkerboard reversal stimuli has been compared with that of other VEP types. It is generally agreed that alternating checkerboard stimuli are much more effective than stimulation with diffuse light flashes. Stimulation with large check sizes may be less effective than stimulation with a small foveal light spot. Measurement of the critical frequency of photic driving to diffuse light flashes may complement the recording of the checkerboard reversal VEP and, in one study, was even more sensitive. Another study found VEPs to a revers-

ing sine wave grating pattern more effective than VEPs to checkerboard pattern shift. VEPs to pattern reversal are as effective as those to pattern shift.[52]

Comparisons of alternating checkerboard VEPs with other EPs have given diverse results: VEPs were more effective than SEPs in all cases, in definite and probable cases, or in early cases of MS; SEPs were more effective than alternating checkerboard VEPs in other studies, especially in the important diagnostic categories of probable and possible MS; cervical SEPs remain more stable than checkerboard VEPs between relapses. These comparisons depend on whether

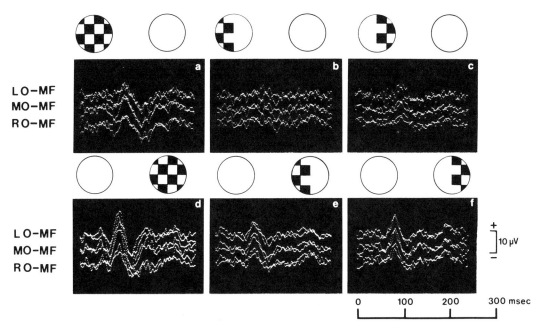

FIGURE 8.2. *Midline and lateral occipital VEPs to monocular full-field and half-field stimulation with checkerboards in a patient with probable MS. Stimulation of the right eye (bottom) produces normal VEPs. Stimulation of the left eye (top) produces full-field VEPs (a) with abnormally prolonged latency and with slightly lower amplitude and longer duration than seen in the full-field VEPs to stimulation of the right eye (d). Stimulation of the left half-field of the left eye produces no peaks that clearly exceed the noise level (b), and stimulation of the right half-field of the left eye produces only low-amplitude peaks of prolonged latency (c). This indicates a conduction defect in the left optic nerve. Stimulation with checkerboard patterns of squares of 30' (a, d) and 60' (b, c, e, f) sidelength. Recording between left occipital (LO), midoccipital (MO), right occipital (RO), and midfrontal (MF) electrodes. Positivity at the occiput is plotted upward. This 38-year-old man has a history of two bouts of transverse myelitis and residual spastic paraparesis. Normal myelogram; no history, systems, or signs suggesting involvement of the visual system. The finding of abnormal VEPs in this patient makes the diagnosis of MS likely.*

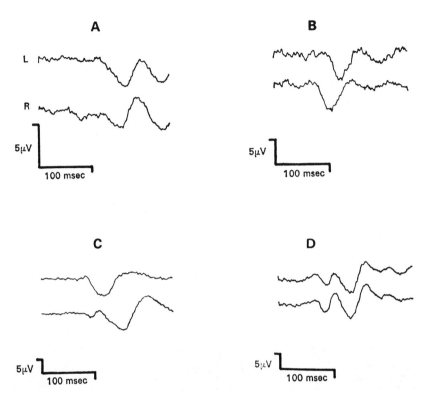

**FIGURE 8.3.** *Transient midline occipital VEPs to full-field checkerboard pattern shift stimulation of the left (L) and right (R) eye in four patients (A ,B, C, D) with MS. (A) The latency of the P$\overline{100}$ is abnormally increased to a similar degree on stimulation of either eye. (B) The P$\overline{100}$ is delayed on stimulation of the left eye only. (C) The P$\overline{100}$ is delayed on stimulation of either eye, much more on stimulation of the right eye. (D) The P$\overline{100}$ and earlier peaks show similar delays on stimulation of either eye. Recording between mid-occipital and vertex electrodes. Negativity at the occiput is plotted upward; the P$\overline{100}$ points down. From Asselman et al. (Brain 98:261, 1975) with permission of the authors and Oxford University Press.*

SEP studies use only arm stimulation or also leg stimulation. BAEPs are generally less effective than checkerboard VEPs in detecting lesions in most diagnostic categories of MS.

### 8.3.1.3 Tumors

Tumors compressing the optic nerve may reduce the amplitude and distort the shape of the VEP (Figure 8.4); in the extreme, they may completely abolish the VEP. Although latency may also be increased, delays of longer than 30 msec, such as can be seen in MS, are rare in tumors. VEP testing is more often used to confirm than to detect optic nerve involvement by

tumors: Patients with abnormal VEPs due to tumors usually have abnormal vision, optic atrophy, or visual field defects indicating an optic nerve lesion. Papilledema due to tumors not directly compressing the optic nerve usually does not cause abnormal VEPs.

### 8.3.1.4 Ischemic optic neuropathy

Ischemic neuropathy of the optic nerve often reduces VEP amplitude. As in tumors, latency may also be increased, but this VEP abnormality is not as characteristic and as pronounced as in MS. Patients with carotid occlusion have been reported to show blurring of vision and reduc-

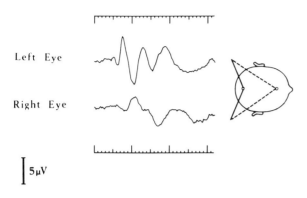

Left  Eye

Right  Eye

5 µV

FIGURE 8.4.   *Transient pattern shift VEPs in a patient with right optic nerve compression due to a sphenoid wing meningioma. Stimulation of the left eye with alternating checkerboard patterns produces a normal midline VEP (top). The VEP to stimulation of the right eye is distorted and delayed (bottom). Time scale 10, 50, and 100 msec. Recording between midoccipital and midfrontal electrodes. Occipital negativity is plotted upward; the P$\overline{100}$ points down. From Halliday et al. (Brain 99:357, 1976), with permission of the authors and Oxford University Press.*

tion of VEP amplitude on exposure to bright light.

### 8.3.1.5 Friedreich's ataxia, hereditary cerebellar ataxia, and hereditary spastic ataxia

The majority of patients with Friedreich's ataxia show a bilateral and fairly symmetrical increase of P$\overline{100}$ latency; amplitude may be reduced more often than in MS, especially in patients with severe visual impairment (Figure 8.5). Studies of other kinds of ataxia showed abnormal VEPs in several patients with hereditary cerebellar ataxia and in one of two patients with olivopontocerebellar degeneration.[41]

### 8.3.1.6 Charcot-Marie-Tooth disease

The VEP was delayed in some patients, even in the absence of clinically apparent optic nerve lesions.[8]

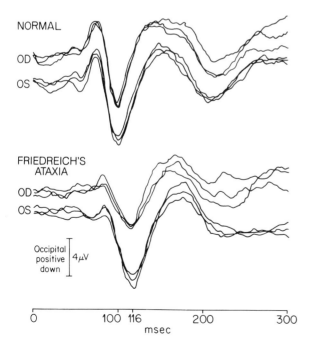

NORMAL

OD

OS

FRIEDREICH'S ATAXIA

OD

OS

Occipital positive down | 4 µV

0      100  116      200      300
msec

FIGURE 8.5.   *Checkerboard pattern shift VEPs in a normal subject (top two tracings) and in a patient with Friedreich's ataxia (bottom two tracings). The patient's VEPs to stimulation of either eye show abnormally delayed P$\overline{100}$ peaks. Monocular full-field stimulation of the right (OD) and left (OS) eye. Recording between midoccipital and vertex electrodes. Negativity at the occiput is plotted upward; the P100 points down. Three tracings are superimposed for each condition. From Nuwer et al. (Ann Neurol 13:20, 1983) with permission of the authors and Little, Brown and Company.*

### 8.3.1.7 Neurosyphilis

VEP latency was found to be increased in one fifth of patients with neurosyphilis, most often those with tabes dorsalis; the VEP was not more useful for diagnosis than other neuro-ophthalmological tests.[23]

### 8.3.1.8 Hereditary spastic paraplegia

VEP latency to pattern reversal stimulus is increased in some patients, but results differ among studies.

### 8.3.1.9 Leber's hereditary optic neuropathy

Leber's disease increases the latency, decreases the amplitude, and eventually abolishes VEPs, in keeping with optic nerve involvement and decreasing vision; changes in VEP shape have also been reported. Asymptomatic family members show little or no VEP abnormalities.

### 8.3.1.10 Traumatic optic nerve lesions

Head injuries may acutely reduce the VEP amplitude suggesting indirect optic nerve damage.

### 8.3.1.11 Leukodystrophies

Pelizaeus-Merzbacher disease, adrenoleukodystrophy, and metachromatic leukodystrophy abolish or delay VEPs.[3,44] The latency may be increased in subclinical cases of adrenoleukodystrophy.

### 8.3.1.12 Chronic renal failure

The VEP may be delayed, especially in patients on chronic hemodialysis.[21]

### 8.3.1.13 Endocrine orbitopathy

The VEP may be delayed even in patients with preserved visual acuity, presumably due to demyelination of the optic nerve.

### 8.3.1.14 Pernicious anemia

Subacute combined degeneration of the spinal cord due to pernicious anemia was associated with slightly delayed VEPs in a few cases.

### 8.3.1.15 Sarcoidosis

VEP latencies may be abnormal even in patients without clinically evident brain or eye involvement.

### 8.3.1.16 Diabetes

Abnormally delayed VEPs have been reported in visually unimpaired patients with diabetes.

### 8.3.1.17 Alcoholism

VEP latency may be increased in a small number of chronic alcoholics and in many patients with Korsakoff's psychosis.

### 8.3.1.18 Phenylketonuria

A small number of patients were found to have increased VEP latency.

### 8.3.1.19 Parainfectious optic neuritis

Optic neuritis after viral diseases may initially abolish VEPs and later be associated with delayed VEPs.

### 8.3.1.20 Postconcussion syndrome

The $\overline{P100}$ latency may be increased after head injuries, especially after those leading to loss of consciousness.

### 8.3.1.21 Peripheral neuropathy

Patients with demyelinating neuropathy due to Guillain-Barré syndrome and chronic inflammatory demyelinating polyneuropathy (CIDP) occasionally had abnormalities on VEP even in the absence of clinical optic nerve dysfunction.[65,87]

### 8.3.1.22 Hysterical blindness

Normal VEPs are found in most patients with hysterical blindness, however, the VEP may be degraded by intentionally poor fixation.

## 8.3.2 Chiasmal Lesions

Pituitary tumors, craniopharyngiomas, and other lesions near the sella turcica compressing

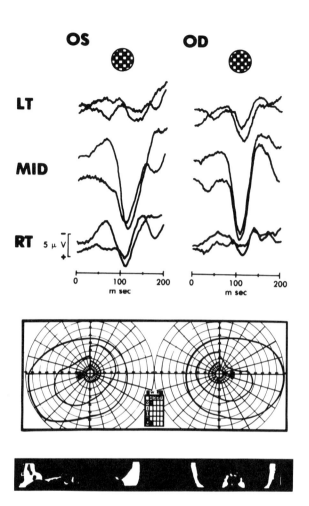

FIGURE 8.6. *Transient checkerboard reversal VEPs (top) in a patient with bitemporal hemianopsia (middle) due to a large intrasellar and suprasellar pituitary tumor (bottom). Full-field stimulation of the left eye (OS) and right eye (OD) and recording with midline (MID) and left (LT) and right (RT) occipital electrodes referred to a midfrontal electrode produce transient VEPs with maximum P$\overline{100}$ amplitude opposite to the stimulated eye; this crossfield asymmetry results from stimulation of the intact nasal half-field of each eye. Occipital negativity is plotted upward; the P$\overline{100}$ points down. From Maitland et al. (Neurology 32:986, 1982) with permission of the authors and Modern Medicine Publications, Inc.*

the optic chiasm produce abnormalities of the monocular full-field VEP that often differ for stimulation of each eye. The abnormalities include reductions of amplitude and gross distortions of waveform, but increased latency may also occur; lateral occipital recordings may show marked asymmetries (Figure 8.6) suggesting the site of the abnormality.

Monocular half-field stimulation in conjunction with midline occipital recordings may produce abnormal VEPs on stimulation of the temporal visual half-field. Even monocular half-field stimulation combined with lateral occipital recordings (Figure 8.7), although the most powerful chiasmal strategy, shows that VEPs are only of limited value in detecting chiasmal lesions even in cases that show visual field

defects. The abnormal crossing of optic nerve fibers in albinism produces asymmetries of the full-field and half-field VEPs.

### 8.3.3 Retrochiasmal Lesions

Tumors, infarcts, and other lesions of the occipital and parietal lobes are difficult to detect with VEPs to full-field stimulation even when they can be detected by visual field defects and imaging studies. As expected, bilateral hemispheric lesions may cause abnormal VEPs to stimulation of either eye. Diffuse cerebral involvement in Alzheimer's disease may increase latency and amplitude of late peaks. Huntington's chorea was found to reduce the amplitude of monocular VEPs in patients and some offspring in one

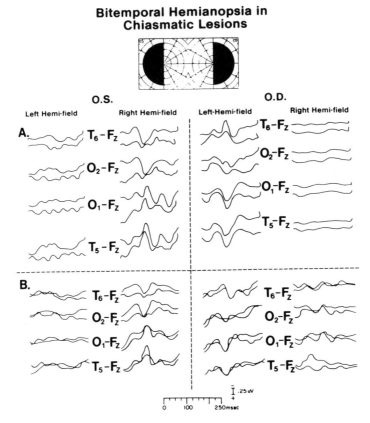

**Bitemporal Hemianopsia in Chiasmatic Lesions**

FIGURE 8.7. *Transient checkerboard reversal VEPs in two patients (A) and (B) with bitemporal hemianopsia due to chiasmal lesions. Stimulation of the left and right half-fields of the left eye (OS) and right eye (OD) produces VEPs when the nasal half-fields are stimulated. Stimulation of the temporal half-fields produces no VEPs in patient A; in patient B stimulation of the temporal half-field of the right eye produces questionable VEPs, and stimulation of the left eye produces no VEPs. Recordings between occipital, temporal, and mid-frontal electrodes as indicated by appreviations of the International 10–20 System for EEG electrode placement. Negativity at the occiput is plotted upward; the P100 points down. From Haimovic and Pedley (EEG Clin Neurophysiol 54:121, 1982) with permission of the authors and Elsevier Scientific Publishers Ireland Ltd.*

study, but produced no VEP changes in another investigation.[32,61] In Parkinson's disease, the VEP amplitude has been reported to be decreased or slightly increased; an increase of latency was found in only one study and was thought to depend on low contrast stimuli. In myotonic dystrophy, abnormalities of latency and amplitude were found in many patients without significant ocular involvement and were thought to be due to prechiasmal and retrochiasmal lesions. VEPs were normal in Tourette's syndrome. Infantile neuraxonal dystrophy with myoclonus epilepsy was associated with increased VEP amplitude. VEP latency was shortened in some cases of photosensitive epilepsy. A binocular increase of VEP latency was found in a patient with palinopsia and in a patient with pseudotumor cerebri. However,

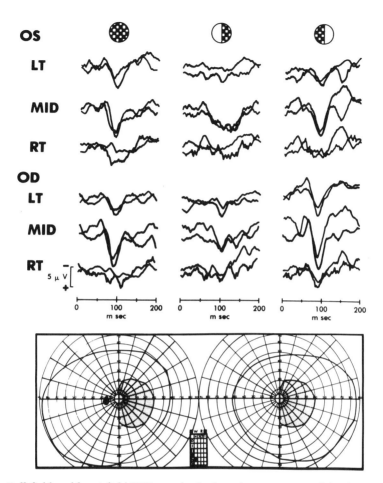

FIGURE 8.8. *Full-field and hemi-field VEPs to checkerboard pattern reversal (top) in a patient with right homonymous scotomas for form and color (bottom) due to a left occipital defect. Stimulation of the left eye (upper three rows of tracings) and of the right eye (lower three rows of tracings) with full-field (left column), right hemi-field (middle column), and left hemi-field (right column) stimuli produces transient VEPs at midline (MID), left (LT), and right (RT) occipital electrodes referred to a midfrontal electrode. Full-field and left hemi-field VEPs are similar and slightly larger ipsilateral to the intact hemi-field. Stimulation of the defective right hemi-field elicits hardly any VEPs. Two samples are superimposed for each VEP. Occipital negativity is plotted upward; the P$\overline{100}$ points down. From Maitland et al. (Neurology 32:986, 1982) with permission of the authors and Modern Medicine Publications, Inc.*

bilateral occipital infarcts leading to cortical blindness, although not usually tested with patterned light stimuli, do not necessarily eliminate the VEP to checkerboard patterns.

The use of binocular visual half-field stimulation and midline occipital recording has shown abnormal VEPs in cases of occipitoparietal tumors and infarcts. The potentially more powerful technique of half-field stimulation in conjunction with bilateral occipital recordings has so far demonstrated its validity mainly in cases of occipital lobectomies, and in patients with lesions of the occipital cortex or optic radiations who also had massive visual field defects (Figure 8.8); even here, the reliability and sensitivity of this method needs further investigation.

## 8.4 OPHTHALMOLOGICAL DISORDERS THAT CAUSE ABNORMAL PATTERN REVERSAL VEPS

Because VEPs depend on sharpness of focus and on check size, they may be used to investigate conditions interfering with the sharp projection of small checks onto the retina. In general, ocular disorders may reduce VEP amplitude but they increase latency only little or not at all.

### 8.4.1 Visual Acuity

The smallest check size giving a VEP may be used to measure the limits of resolution.

### 8.4.2 Refractive Errors

At normal visual acuity, maximum VEP amplitude is obtained with checks of 10–15′. Refractive errors increase the size of the most effective checks. The magnitude of the error is indicated by the corrective lenses required to reverse this effect.

### 8.4.3 Amblyopia Ex Anopia

The amplitude of transient VEPs to small check sizes is reduced; the latency may be increased. Although this finding suggests a prechiasmal conduction defect, it is not easily mistaken for an optic nerve lesion because it is characterized by a history dating from childhood. Amblyopia had been studied more extensively with steady-state VEPs to alternating checkerboard patterns and to sine wave gratings.

### 8.4.4 Nutritional and Toxic Amblyopia

Most patients with nutritional amblyopia who showed defective color vision and central or cecocentral scotomata had delayed VEPs in one study, but not in another. VEP amplitude may be reduced symmetrically. Quinine amblyopia may cause asymmetrical VEP abnormalities. Increased latency may reveal subclinical optic neuropathy during ethambutol treatment.

### 8.4.5 Retinopathy

Idiopathic central serous retinopathy may increase the latency and decrease the amplitude of the VEP of the involved eye.

### 8.4.6 Glaucoma

The latency of the VEP may be slightly increased as the result of glaucomatous optic nerve damage associated with field defects.

### 8.4.7 Corneal Opacity, Meiosis, and Cataract

A significant reduction of stimulus luminance reduces VEP amplitude and may increase VEP latency slightly.

### 8.4.8 Congenital Oculomotor Apraxia

Normal visual acuity has been demonstrated with VEPs.

# 9

# The Transient VEP to Diffuse Light Stimuli

## 9.1 CLINICAL USEFULNESS OF THE FLASH VEPS

VEPs to diffuse light flashes, although widely used in early clinical VEP studies, have now been largely replaced by VEPs to patterned light except:

- in testing infants, children, and other subjects who do not reliably focus on pattern stimuli or who can not clearly see the stimulus pattern, even if it contains large pattern elements;
- for the study of patients who show no responses to pattern light stimuli; and
- for determining the critical frequency of photic driving.

Because reports of abnormal flash VEPs in other disorders have lost much of their clinical importance, they are discussed only briefly in this book.

## 9.2 THE NORMAL TRANSIENT FLASH VEP

Reports on normal VEPs to diffuse light flashes usually describe up to seven peaks (Figure 9.1)

but differ widely with regard to peak polarity, latency, and amplitude. These differences are partly due to different stimulating and recording methods which alone would not necessarily reduce the value of the flash VEP as a diagnostic tool as long as stimulus and recording conditions were maintained rigidly constant in the testing laboratory. However, even VEPs recorded in the same laboratory vary so much among subjects, and in the same subject with time, that it is very difficult to define a standard normal VEP as having a certain number of peaks of constant polarity, latency, and amplitude unless VEPs from many subjects are averaged together. For diagnostic purposes, each laboratory may select one or a few of the least variable VEP parameters for use in clinical testing. Usually, these are the latency of the major positive peak, occurring some time between 50 and 100 msec after the flash, and the latency of the subsequent, large negative peak, occurring between 100 and 250 msec after the flash; a negative wave preceding the major positive wave may also be suitable.

Peaks of shorter and longer latency than 50–250 msec are found even less constantly.

Early peaks of low amplitude may be recorded in some normal subjects with the usual recording methods. Recordings with narrow filter settings may pick up short-latency wavelets, some of which probably represent excitation of subcortical segments of the visual pathway, but these potentials have not so far been put to clinical use in the same manner as the far-field recordings commonly used in AEP and SEP recordings. Late peaks of the cortical VEP to diffuse light flashes are seen often and may have rather large amplitude but are more variable than those between 50 and 100 msec. The early and late peaks of the VEP may be followed by a train of rhythmical waves at or near the frequency of the alpha rhythm (Figure 9.1). This rhythmic after-discharge varies enormously and has no known clinical importance.

## 9.3 SUBJECT VARIABLES

### 9.3.1 Age

#### 9.3.1.1 Premature infants

The flash VEP after 24 weeks of gestation consists of an occipitally negative peak at 200–300 msec. At 32–35 weeks, a positive peak of less than 200 msec appears before the negative peak. With further development, the amplitude of the positive peak increases, that of the negative peak decreases, and the latencies of both peaks decrease. The features of these VEPs differ in the different sleep stages of the premature infant. With increasing gestational age, the distribution of the VEP expands from the occipital to the frontocentral areas.

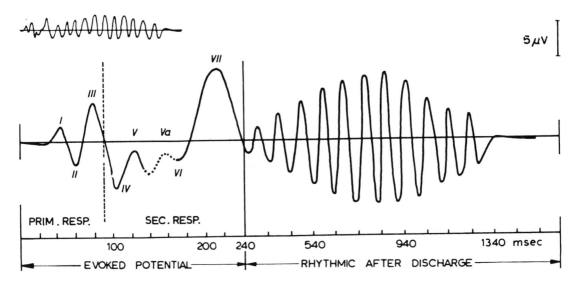

FIGURE 9.1.   *Diagram of the normal VEP to diffuse light flashes. This tracing is the grand average of VEPs from 75 subjects. The VEP is here divided into a primary response (prim. resp.) and a secondary response (sec. resp.) and is followed by a rhythmical afterdischarge. Note the change in time scale between VEP and afterdischarge. The inset in the left upper corner shows VEP and afterdischarge on the same time scale. Recording between midoccipital and midparietal electrodes. Negativity at the occipital electrode is plotted upward. From Cigánek (EEG Clin Neurophysiol 13:165, 1961) with permission of the author and Elsevier Scientific Publishers Ireland Ltd.*

### 9.3.1.2 From full-term to adult age

At term, the flash VEP often consists of a sequence of negative-positive-negative peaks. Later, additional peaks appear and peak latencies decrease. The latencies of the early peaks reach adult values in early childhood; later peaks do not completely mature until puberty.[9] Latency and amplitude of VEPs vary even more in children than in adults. Flash VEPs in children therefore do not reliably indicate maturation and are even less sensitive in detecting lesions than flash VEPs in adults. Longitudinal studies, if practical, may be more rewarding.

### 9.3.1.3 Old age

The latency of VEP peaks increases after the age of 65.[30] The amplitude may increase or decrease.

## 9.3.2 Pupillary Size

An increase of pupillary size acts like an increase of stimulus intensity. An artificial pupil or other device is needed to keep retinal illumination by stimulus and background light constant throughout the recording. The subject's eyes should be open during stimulation.

## 9.4 STIMULUS CHARACTERISTICS

## 9.4.1 Stimulators

### 9.4.1.1 Xenon flash tubes

Stroboscopic flash stimulators, found in most EEG laboratories, are the most commonly used source of diffuse light flashes in EP laboratories. They give brief and very intense flashes at rates up to more than 100/sec. The lamp may be directed at the eyes of the subject, or at a reflecting or transparent screen in front of the subject's eyes; light coming from a screen is more likely to illuminate parts of the retina evenly and to avoid stimulation by the pattern represented by the lamp.

Two features of flash tubes may have clinical importance: (1) the light intensity may decrease at higher stimulus rates so that VEPs at different rates may not be directly comparable; (2) the flashes may be accompanied by a click that may act as an auditory stimulus and should be eliminated by acoustic shielding.

### 9.4.1.2. Electromechanical stimulators

The beam of a constant light source is brought into a narrow focus where it can be interrupted by an electronically controlled shutter producing brief flashes or sustained light pulses.

### 9.4.1.3 Fluorescent lamps

Fluorescent tubes that can be driven by direct current may be used to give light pulses of different length and shape; these lamps may be placed behind a diffuser to give uniform illumination.

### 9.4.1.4 Light-emitting diodes

Light-emitting diodes (LED) may be used for flash stimulation and may be mounted in goggles or contact lenses for VEP studies during surgery or for testing of infants, small children, and patients in coma.

### 9.4.1.5 Full-field (Ganzfeld) flash stimulators

Brief flashes of specified luminance are delivered into a reflecting sphere into which the subject looks through a round cutout. This arrangement provides illumination of the entire visual field independent of the direction of gaze; distance and intensity of the stimulus and background light can be controlled easily. Ganzfeld stimulation is used for quantitative studies, especially for the ERG.

## 9.4.2 Stimulus Rate

Because VEPs to diffuse light stimuli may last 500 msec, stimulus rates should not exceed 2/sec for transient VEPs.

## 9.4.3 Stimulus Intensity

Most VEP peaks vary with the stimulus intensity: Latency decreases and amplitude increases with increasing intensity to a saturation level that depends on many other parameters such as monocular versus binocular stimulation. The distance between the flash lamp or the screen used to present flash stimuli and the subject's eye should be 30–45 cm and must be kept constant in each laboratory. Moderate ambient light is usually present and should be rigidly controlled because it alters the luminance change produced by the stimulus.

## 9.5 RECORDING PARAMETERS: ELECTRODE PLACEMENTS AND COMBINATIONS

Routine recording electrodes may be placed in midoccipital and left and right occipital positions as for recordings of checkerboard pattern reversal VEPs. Each of these electrodes may be referred to interconnected ear electrodes for three channels of recording. A fourth channel may be added to record VEPs between a vertex electrode and the interconnected ear electrodes or to record the ERG.

The distribution of the flash VEP has been studied extensively. A few points are of practical importance. Early peaks have a maximum over the occipital areas whereas some later peaks are most prominent over the vertex. Latency shows little asymmetry, but amplitude may differ markedly in normal subjects, and a VEP may be completely absent over one side of the head in up to 5% of normal subjects. Occipital VEPs recorded from the scalp are similar to VEPs recorded from the calcarine cortex. Artifacts from ERG and muscle contractions may contaminate scalp recordings in some areas.

The VEP is due to stimulation of the fovea. It is difficult to evaluate the relationship between other retinal stimulus sites and VEP distribution precisely. Restricted flash stimuli scatter in the optic media of the eye and reach wide parts of the retina. Diffuse light stimuli are therefore not of much value for half-field stimulation. Very small light spots act as a patterned stimulus.

## 9.6 CRITERIA DISTINGUISHING ABNORMAL FLASH VEPS

Because of the great variability of flash VEPs, the only reliable criterion of abnormality is the complete absence of the monocular flash VEP unexplained by technical problems. Some other criteria may suggest possible abnormalities. In most laboratories, the latencies to one or more of the major positive or negative peaks are chosen. Different normative values should be determined for different ages, especially for the young age groups in which patterned stimuli cannot be used.

## 9.7 NEUROLOGICAL DISORDERS THAT CAUSE ABNORMAL FLASH VEPS

The division into prechiasmal, chiasmal, and retrochiasmal is satisfactory except for disorders that involve the brain widely; these are arbitrarily classified depending on the site of involvement most frequently studied with VEPs.

### 9.7.1 Prechiasmal Lesions

#### 9.7.1.1 Multiple sclerosis

VEPs to flash stimuli may be used in patients with impaired vision who can not see patterned stimuli. The preservation of flash VEP in the absence of pattern VEPs indicates that at least some fiber connections between retina and occiput are preserved. The flash VEP has been described to have abnormally increased latency, decreased amplitude, and deformed shape in many patients with MS. In patients who can see checkerboard patterns, VEPs to this stimulus have much greater diagnostic power than VEPs to flash stimuli.

#### 9.7.1.2 Surgical monitoring of optic nerve compression

The flash VEP has been used to monitor compression and decompression of the optic nerve during orbital and intracranial surgery. However, the VEP varies with the level of general anesthesia and has been found to be an unreliable indicator of optic nerve function during surgery.

### 9.7.1.3 Other prechiasmal lesions

Abnormal flash VEPs have been described in many conditions, some of which may also cause retrochiasmal lesions:

- tumors compressing the optic nerve or chiasm
- ischemic optic neuropathy
- Friedreich's ataxia
- hereditary spastic paraplegia
- lipid storage diseases
- leukodystrophies
- Leber's optic neuropathy
- toxic optic neuropathy

## 9.7.2 Retrochiasmal Lesions

### 9.7.2.1 Head injury, increased intracranial pressure, and hydrocephalus

Abnormal VEPs, AEPs, and SEPs, seen in many patients in coma after severe head injuries, have been found helpful in diagnosing focal deficits and in predicting general outcome. Head trauma may cause abnormal VEPs due to optic nerve injury. In postconcussion syndrome, VEP abnormalities have been reported to identify patients with organic symptoms.

Increased intracranial pressure due to severe head injury or to hydrocephalus with shunt malfunction may increase VEP latency. Removal of spinal fluid in hydrocephalus may decrease VEP latency.

### 9.7.2.2 Anoxic encephalopathy and brain death

In acute cerebral anoxia, the VEP generally deteriorates and disappears earlier than the EEG and ERG, but may be preserved in the absence of EEG activity. Although VEPs are absent in brain death, their absence does not prove brain death: The VEP seems to reflect the condition of

the hemispheres more than that of the brainstem centers essential for survival.

### 9.7.2.3 Cortical blindness

The bilateral occipital lesions causing cortical blindness abolish occipital VEPs, as expected, in many cases, but peaks of normal or prolonged latency may persist at the occiput or at the vertex in some cases. Children with cortical blindness often have preserved VEPs showing some abnormalities. The preservation of VEPs in cortical blindness reduces the ability of the VEP to diagnose blindness, especially in children.

The prognostic value of a preserved VEP to diffuse flashes in cortical blindness is also limited. Even though VEPs often improve with return of visual function, preservation of VEPs and degree of VEP abnormality do not necessarily indicate chances for recovery. Stimulation with diffuse light has been used much more often for the study of cerebral blindness than stimulation with patterned light.

### 9.7.2.4 Seizures

Patients with seizures triggered by flash stimuli (photosensitive, photic, or photogenic seizures), children with generalized seizures, and patients with myoclonus epilepsy have been found to have abnormally large VEPs. Other interictal VEP abnormalities are inconsistent.

VEPs may be preserved during 3/sec spike-wave discharges. Seizures induced by unilateral electroconvulsive treatment depress VEPs on the treated side for 15 minutes postictally.

### 9.7.2.5 Other retrochiasmal disorders

Abnormal VEPs have been studied in many conditions, some of which also cause prechiasmal problems:

- tumors and strokes
- Alzheimer's disease
- Jakob-Creutzfeldt disease
- Parkinson's disease
- Wilson's disease

- Huntington's chorea
- renal failure and hemodialysis
- Down's syndrome
- phenylketonuria
- hyperglycinemia
- Menke's "kinky hair disease"
- hyperthyroidism
- hypothyroidism
- alcoholism
- smoking tobacco
- developmental and perinatal disorders
- learning disorders and dyslexia
- hyperactive children on treatment with methylphenidate
- psychiatric disorders

## 9.8 OPHTHALMOLOGICAL DISORDERS THAT CAUSE ABNORMAL FLASH VEPS

Flash VEPs may be of some value in ocular disorders that cause impaired vision precluding the use of patterned stimuli. Some retinal lesions cause ERG abnormalities in addition to VEP abnormalities and may be investigated by combined ERG and VEP recordings.

### 9.8.1 Increased Intraocular Pressure and Glaucoma

Flash stimulation of glaucomatous eyes elicits VEPs of abnormally low amplitude, especially at low stimulus luminances. These VEPs are very sensitive to increased intraocular pressure. Experiments raising the intraocular pressure of normal eyes show that the VEP disappears at the point of pressure blinding, presumably because of retinal ischemia. The VEP is more sensitive than the ERG, which persists even when the intraocular pressure briefly rises above the blood pressure of the ophthalmic artery, suggesting that the retina is less sensitive to pressure than the prelaminar part of the optic nerve.

### 9.8.2 Injury to the Eye or Optic Nerve

Eye injury and head trauma with indirect optic nerve injury may decrease the amplitude and increase the latency of the flash VEP.

### 9.8.3 Retinal Detachment

The flash VEP may be preserved even in cases of severely abnormal ERG and completely detached retina.

### 9.8.4 Retinitis Pigmentosa

The flash VEP is usually normal or less reduced than the ERG.

### 9.8.5 Amblyopia

The VEP to diffuse light is usually normal in amblyopia ex anopia. Toxic amblyopia due to alcohol and tobacco or to quinine has been reported to cause VEP abnormalities.

### 9.8.6 Cataract

The flash VEP has been said to be of some help in predicting visual function after cataract surgery.

### 9.8.7 Other Ophthalmological Disorders

Abnormal flash VEPs have been described in the following conditions:

- central retinal artery occlusion
- optic nerve hypoplasia
- optic nerve head drusen
- congenital nystagmus

# 10

# VEPs to Other Stimuli

## 10.1 TRANSIENT VEPS TO CHECKERBOARD APPEARANCE, DISAPPEARNACE, AND FLASH

The VEP to pattern appearance, or onset, differs from that to pattern disappearance, or offset; both differ from the VEP to check reversal (Figure 10.1). The VEP to pattern offset is usually more similar to the VEP to pattern reversal than is the VEP to pattern onset, although this may vary with stimulus conditions. VEPs to pattern onset differ depending on whether the onset is associated with an increase, decrease, or no change of overall luminance.

When pattern presentation is shortened to about 30 msec, separate VEPs to onset and offset can no longer be distinguished; VEPs to such brief flashes of patterned light are the compound effect of onset and end of pattern presentation. These VEPs differ in shape and distribution from VEPs to other checkerboard presentations.

Maturation of VEPs to checkerboard onset and flash resembles that of VEPs to checkerboard alternation in that both the peak latency and the effective check size decrease during infancy and childhood in a manner suggesting that spatial resolution improves rapidly to the age of six months and more slowly to puberty. Amplitude is said to be higher at young and old than at middle age and to depend on luminance in a manner changing with age.

Neurological studies have shown VEPs to pattern onset or flash to be abnormal in MS, hemispherectomy, phenylketonuria, and schizophrenia. Occasional comparisons with VEPs to pattern reversal have suggested that VEPs to pattern onset may be more effective in the diagnosis of optic nerve lesions in MS and in the study of abnormal fiber crossing in human albinism.

Ophthalmological investigations have used these VEPs to study refraction, amblyopia ex anopia, strabismic amblyopia, cataracts, and binocular vision.

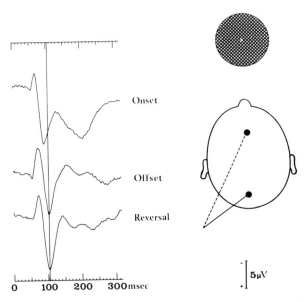

FIGURE 10.1.  *Normal transient VEPs to onset, offset, and reversal of a checkerboard pattern. Each tracing is the grand average of recordings from 10 subjects. Right monocular stimulation with 50' checks in a 32° round field. Recording between midoccipital and midfrontal electrodes. Occipital negativity is plotted upward; the P100 points down. From Kriss and Halliday (In Evoked potentials: Proceedings of an international evoked potentials symposium held in Nottingham, England, ed. C. Barber, pp. 205–212, MTP Press, 1980) with permission of the authors and MTP Press Lancaster.*

## 10.2 STEADY-STATE VEPS TO CHECKERBOARD PATTERN REVERSAL, APPEARANCE, DISAPPEARANCE, AND FLASH

### 10.2.1 The Normal Steady-State Pattern VEP

Rhythmical VEPs to checkerboard pattern stimulation may be produced at stimulus rates beginning at about 4/sec; these steady-state VEPs are often greatest at 6–8/sec and smaller at higher rates (Figure 10.2). Like transient VEPs, steady-state VEPs are largest with small check sizes. Larger check sizes are more effective at higher stimulus rates.

Steady-state VEPs to patterned stimuli, like those to diffuse light, are largely explained by superimposition of transient VEPs,[48] but have not been thought to reflect different aspects of visual function than do transient VEPs. Amplitude is the most commonly measured parameter of steady-state VEPs. Because the latency of

steady-state VEPs can not be measured directly, other methods must be used to characterize their timing.

### 10.2.2 Maturation

The check size producing the largest steady-state VEP to patterned flashes decreases during the first few months of life. By six months of age, reversing checks of 7.5' and 15' produce the largest response, as they do in adults with 20/20 visual acuity. VEPs to faster stimulation with small sizes mature later than VEPs to slower stimulation.

### 10.2.3 Neurologic Disorders that Cause Abnormal Steady-State Pattern VEPs

Steady-state checkerboard VEPs are used only rarely in clinical diagnosis. Abnormalities of amplitude and timing have been reported in retrobulbar neuritis (Figure 10.3), MS, traumatic neuritis, and chiasmal and retrochiasmal

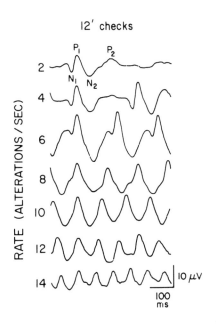

FIGURE 10.2. *Development of the steady-state VEP to checker-board pattern reversal with increasing stimulus rate; the rate, in pattern reversals per second, is indicated at the left of the tracings. At stimulus rates of 2/sec and 4/sec, transient VEPs with peaks $N_1$, $P_1$, $N_2$, and $P_2$ can be distinguished. At a rate of 6/sec, VEPs overlap. At 8/sec and above, individual VEP features give way to the rhythmical steady-state VEP consisting of waves at the rate of stimulus. Occipital negativity is plotted down; the $P\overline{100}$, here named $P_1$, points up. From Sokol (In Electrodiagnosis in clinical neurology, ed. M. J. Aminoff, pp. 348–369, Churchill Livingstone) with permission of the author and Churchill Livingstone Medical Division of Longman Group Ltd.*

tumors and infarcts presenting with visual field defects. Study of steady-state VEPs in addition to transient VEPs may increase the diagnostic yield of retrochiasmal VEP studies. Steady-state checkerboard VEP may be preserved in cortical blindness due to bilateral occipital lesions.

### 10.2.4 Ophthalmological Disorders

#### 10.2.4.1 Refraction and accommodation

The amplitude of steady-state VEPs to patterned stimuli may be used to determine the corrective lenses that result in the largest VEP amplitude, indicating optimum visual acuity. The same method has been applied to measure accommodation. Estimates of visual acuity have been made with a method using narrow-band filtering of steady-state responses to changing checkerboard sizes and with a method using spectral analysis of occipital responses instead of averaging.

#### 10.2.4.2 Amblyopia

The decreased visual acuity in amblyopia ex anopia is manifested by an increase of the check size producing VEPs of the largest amplitude, suggesting that the VEP of the amblyopic eye is due mainly to parafoveal elements of the retina; the phase relation between peaks of VEPs to stimulation of each eye may be changed if binocular vision is not preserved. Toxic amblyopia may produce interocular amplitude differences.

#### 10.2.4.3 Macular lesions

Macular degeneration may reduce the VEP without altering the ERG. A macular cyst was found to increase the effective check size to that which normally stimulates paramacular regions.

#### 10.2.4.4 Glaucoma

The VEPs to stimulation of retinal quadrants in glaucomatous eyes with field defects have been found to have abnormal timing and amplitude; VEPs from eyes with chronic ocular hypertension alone showed no such changes. VEPs of glaucomatous eyes had reduced amplitude if the central 10° of the field were defective; they were found to be very sensitive to acute changes of intraocular pressure.

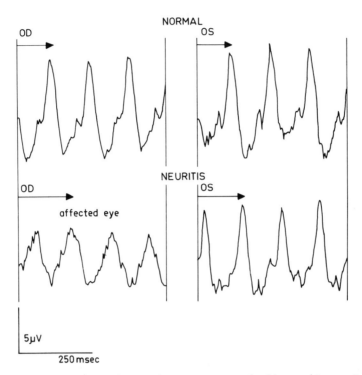

FIGURE 10.3.　*Steady-state checkerboard reversal VEPs in a normal subject and in a patient with right optic neuritis. In the normal subject, the latency from the stimulus at the beginning of each tracing to the first upward deflection (horizontal arrows) is equal for the right (OD) and left (OS) eye. In the patient with optic neuritis, the time interval from the stimulus to the VEP peak is changed, suggesting a marked prolongation for the clinically affected eye and a moderate prolongation for the clinically normal eye. These phase shifts indicate bilateral optic nerve involvement. Stimulation with 8/sec checkerboard pattern reversals. Recording between midoccipital and earlobe electrodes. Occipital positivity is plotted upward. From Wildberger et al. (Ophthal Res 8:179, 1976) with permission of the authors and S. Karger AG.*

## 10.3 STEADY-STATE VEPS TO DIFFUSE LIGHT STIMULI

### 10.3.1 The Normal Steady-State Flash VEP

Repetitive stimulation with light flashes or with sinusoidal modulation of luminance at rates over 5/sec produces a rhythmical wave of the frequency of the repetitive stimulus; harmonic components may be superimposed or predominate (Figure 10.4). The timing of each peak with respect to the preceding stimulus shows a rapid phase shift at a stimulus rate of about 10/sec and a phase lag proportional to the stimulus frequency at higher stimulus rates; it is independent of the frequency of the alpha rhythm of

the subject under study. Like the steady-state VEP to checkerboard pattern stimulation, the steady-state VEP to diffuse light is largely due to superimposition of transient VEPs.

The steady-state VEP amplitude varies with the stimulus frequency, luminance, size, and many other parameters. Under some laboratory conditions, maximum amplitudes were found at a low frequency of about 10/sec, at a medium frequency of 13–25/sec, and at a high frequency of 40–60/sec. The distribution of the steady-state VEP varies depending on stimulus frequency, retinal site of stimulation, and other factors.

Steady-state VEPs can be elicited at high rates depending on stimulus intensity and other factors. The highest rate at which a steady-state VEP can be distinguished is called the *critical frequency of photic driving* (CFPD) and varies inversely with age. It was found to have mean values of 72/sec between the ages of 20 and 30 years, 68/sec between 30 and 60, and 60/sec above 60 years in one laboratory. Steady-state VEPs may be recorded at rates above the critical fusion frequency.

Two problems may arise when steady-state VEPs are used in neurologic diagnosis; both pertain to measuring the response. First, VEP amplitude varies so much among subjects that abnormalities can be detected only by comparing VEPs in the same subject, especially VEPs recorded simultaneously on the two sides of the head or VEPs elicited successively by stimulation of either eye. Second, strict criteria for determining the presence of steady-state VEPs at slightly different frequencies have not yet been developed. Nonetheless, it seems that even rather gross estimates of the presence of VEPs at frequencies differing by as much as 10/sec or more are sufficiently sensitive to indicate abnormalities.

## 10.3.2 Prechiasmal Lesions that Cause Abnormal Steady-State Flash VEPs

Optic neuritis and MS may reduce the amplitude of steady-state VEPs and may change the time relation between stimuli and peaks; however, steady-state VEPs to diffuse light have been found to be diagnostically less useful than VEPs to patterned light stimuli. Studies using Fourier analysis of the steady-state VEP instead of averaging showed that flicker rates of 13–25/sec were most effective in detecting abnormalities in MS.

Determination of the CFPD may increase the diagnostic yield of transient VEPs to alternating checkerboard patterns in MS. It has been claimed that the CFPD may even be more effective in this regard than the transient pattern

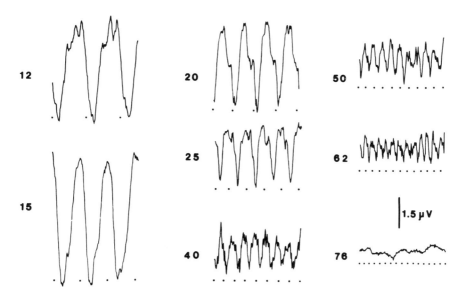

FIGURE 10.4. *Steady-state flash VEPs and critical frequency of photic driving (CFPD) in a normal 21-year-old subject. Each tracing represents a steady-state VEP to diffuse light flashes. Stimulus rate (flashes per second) is indicated to the left of the tracings. Dots below the tracings mark flash stimuli. Steady-state VEPs are elicited with flashes up to 62/sec but not with flashes at 76/sec. Monocular flash stimulation after pupillary dilatation. Midoccipital recording. Each tracing represents 204.8 msec. From Celesia and Daly (Arch Neurol 34:403, 1977) with permission of the authors and the American Medical Association.*

VEP. The steady-state VEP has also been reported to be abnormal in patients with chronic papilledema.

### 10.3.3 Retrochiasmal Lesions that Cause Abnormal Steady-State Flash VEPs

Steady-state VEPs to diffuse light are not commonly used for the study of retrochiasmal lesions. Power spectral analysis of steady-state VEPs may show abnormal VEP distribution in some cases of tumors and infarcts causing homonymous hemianopia. Patients with photosensitive seizures have been found to show changes of VEPs elicited by the beginning of repetitive trains of flashes.

## 10.4 VEPS TO SINE WAVE GRATINGS

### 10.4.1 The Principle of Sine Wave Grating Stimuli

Sine wave grating stimuli consist of alternating light and dark stripes with midlines of maximum and minimum luminance and a gradual transition of luminance between them; precisely, the change of luminance has the form of a sine wave. The size of the pattern elements is measured between two light or dark stripes and expressed as cycles per degree (c/deg) of visual angle. These stripes are used because they contain only a single spatial frequency, namely, that of the sine wave representing the brightness modulation. This differs from patterns with sharp borders between light and dark elements which represent sudden transition of brightness that can be equated with a visual square wave and represent a mixture of visual sine waves of different frequencies. Stimulation with sine wave gratings gives the investigator the opportunity to select one spatial frequency, to study the ability of the visual system to transmit information in discrete channels of spatial frequency; another slight advantage of the sine wave grating stimuli is that, because they do not contain sharp borders, they are less affected than checkerboard patterns by blurring of up to 2 diopters.

### 10.4.2 Stimulators for Sine Wave Gratings

Grating stimuli are usually produced with signal generators on an oscilloscope screen, although a computer driving an oscilloscope or video monitor and optical techniques using polarized light have also been described. Laser beams have been used to form sinusoidal grating patterns directly on the retina.

### 10.4.3 Transient VEPs to Sine Wave Grating Stimuli

#### 10.4.3.1 The normal transient sine wave grating VEP

The transient VEP to the reversal of a sine wave grating pattern and the transient VEP to the appearance of a grating pattern of less than 1 c/deg are similar to the transient checkerboard reversal VEP: They consist mainly of an occipitally positive peak of a latency of about 100 msec; this peak is preceded by a smaller negative peak. Transient VEPs to the appearance of a grating pattern of over 1 c/deg consist of two early and two late negative-positive peaks whose latency and amplitude varies with spatial frequency and contrast depth (Figure 10.5).

#### 10.4.3.2 Neurologic disorders that cause abnormal transient sine wave grating VEPs

10.4.3.2.1 *Multiple sclerosis.* The VEP latency is increased in the majority of patients.[12] Abnormalities are more common with high spatial frequencies. Sine wave gratings may be more effective than checkerboard patterns in detecting optic nerve lesions in MS, and testing with gratings of different spatial orientation may increase the diagnostic yield.

10.4.3.2.2 *Parkinson's disease.* The latency may be increased.

#### 10.4.3.3 Ophthalmological disorders that cause abnormal transient sine wave grating VEPs

Increased VEP latencies have been reported for ocular hypertension and open-angle glaucoma.

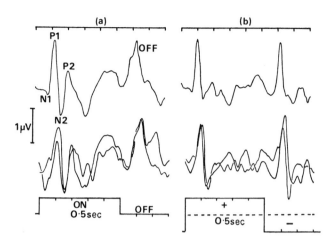

FIGURE 10.5. *Normal transient VEPs to the onset, end, and reversal of a sine wave grating stimulus. Pattern onset, at the upstroke of the marker tracing at the bottom in (a), produces a VEP with two negative and two positive peaks (N1, P1, N2, P2). Pattern offset, at the downstroke of the marker tracing in (a), produces a simpler VEP consisting mainly of a positive deflection. Pattern reversals, at the upstrokes and downstrokes of the marker channel in (b), also produce a VEP consisting mainly of a positive deflection. The lower tracings were obtained after cycloplegia and paralysis of the extraocular muscles which caused no important changes. Monocular stimulation with a grating pattern of 3 c/deg. Recording between midoccipital and temporal electrodes. Occipital positivity is plotted upward. Reprinted from Kulikowski and Leisman, Vision Res 13:2079, 1973, with permission from the authors and Pergamon Press, Ltd.*

## 10.4.4 Steady-State VEPS to Sine Wave Gratings

### 10.4.4.1 The normal steady-state sine grating VEP

Steady-state VEPs to sine wave gratings consist of a rhythm whose amplitude varies with spatial frequency and reaches maximum at a value corresponding with that of the most effective checkerboard pattern (Figure 10.6); the amplitude also increases with contrast, within limits. The curve relating VEP amplitude to spatial frequency has a similar shape and peak as that obtained from measuring subjective contrast sensitivity, that is, the threshold for detecting sine wave gratings of different spatial frequencies at different levels of contrast (Figure 10.6). The measurement of subjective contrast thresholds at different spatial frequencies, or visuogram, gives a more thorough description of visual ability than Snellen's method of measuring visual acuity and has been used in some clinical investigations of

MS and retrobulbar neuritis, other prechiasmal, and retrochiasmal lesions, anisometropic and strabismic amblyopia, and glaucoma; the measurement of dynamic contrast sensitivity, using flickering sine wave grating patterns, seems more effective in detecting abnormalities than the measurement of static contrast sensitivity, using stationary gratings. Although visual impairment usually involves discrimination of high spatial frequencies (i.e., small sizes) before affecting medium and low spatial frequencies (i.e., larger sizes), some patients show selective involvement of discrimination of medium frequencies; this impairment, although causing complaints of blurred vision, escapes detection by conventional visual acuity testing with letters that have sharp black and white interfaces representing mixed spatial frequencies. Another approach to the study of contrast effects uses intermittent changes of contrast depth short of reversal as a VEP stimulus to determine the contrast modulation threshold.

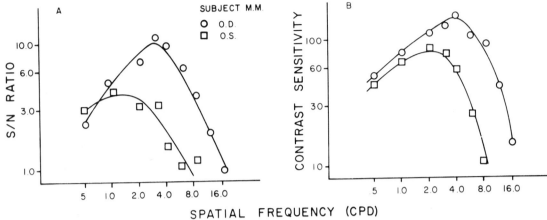

FIGURE 10.6.   *Spatial tuning curves for steady-state VEPs and psychophysical contrast thresholds to sine wave gratings, recorded in a subject with a normal right eye (circles) and an amblyopic left eye (squares). The curves in (A) show the magnitude of the steady-state VEP to sine grating stimuli (signal) and to blank stimulation (noise and by calculating the ratio of the power at the stimulus frequency under both conditions (S/N ratio). Stimulus contrast was 0.44. Recording between midoccipital and ear electrodes. The curves in (B) show the contrast threshold which was determined by having the subject view the same grating stimuli, by increasing or decreasing the contrast at each spatial frequency, by recording the contrast level at which the subject indicates appearance or disappearance of the pattern, and by plotting the reciprocal of that contrast level as contrast sensitivity. Both types of tuning curves show a peak at 3–4 c/deg for the normal eye. For the amblyopic eye, the peaks of both curves lie at a lower frequency, and the VEP tuning curve is less sharply peaked. From Levi and Harwerth (Ophthalmol Visual Sci 17:571, 1978) with permission of the authors and Lippincott/Harper and Row.*

Maturation of contrast sensitivity as studied with sine wave grating VEPs, reaches near adult levels for low and medium spatial frequencies at six months of age. However, maximum contrast sensitivity continues to increase, especially for higher spatial frequencies, to the end of the second or third year.

### 10.4.4.2 Neurologic disorders that cause abnormal steady-state sine wave grating VEPs

The steady-state VEP has been used to detect optic nerve lesions in MS. In cortical blindness, the steady-state VEP may persist.

### 10.4.4.3 Ophthalmological disorders that cause abnormal steady-state sine wave grating VEPs

Amblyopia reduces the amplitude and the effective spatial frequency of steady-state VEPs to sine wave gratings (Figure 10.6). Astigmatic amblyopia is characterized by a reduction of VEP amplitude to gratings oriented in the direction of the reduced visual acuity. Glaucoma preferentially affects VEPs to gratings of low spatial frequency. Patients with ocular opacities have been studied with interference fringes produced with a laser interferometer to assess their visual function.

Sine wave grating patterns have been used to determine visual acuity with a method giving faster results than the averaging VEPs. Steady-state responses are elicited at a fairly high stimulus rate with a grating pattern that gradually changes from low to high spatial frequency. The occipital responses are extracted from the background activity with a filter sharply tuned to the stimulus frequency or with power spectral analysis which locks in on the stimulus frequency and gives a write-out of the response amplitude for the different spatial frequencies,

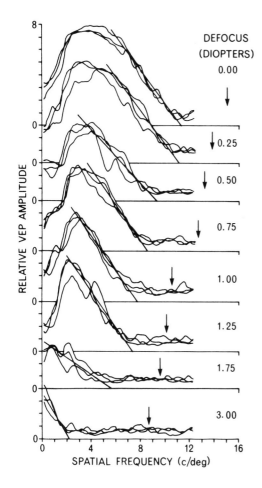

FIGURE 10.7. *Spatial tuning curves obtained with a rapid linear sweep method. Each tracing represents occipital electric responses to fast repetitive stimulation with reversing sine wave grating patterns of continuously increasing spatial frequency. The recordings are not averaged but filtered with a narrow-band filter set to the value of the stimulus rate and reflect response amplitude at the stimulus rate. Three tracings are superimposed for each condition. The intersection of the descending slope of the tracing with the horizontal axis represents VEP acuity; the psychophysically measured visual acuity is indicated by arrows. The effect of defocusing (numbers at the right of the tracings) shows the sensitivity of the method to induce refractive errors and its potential as a means of measuring refractive errors. Recording between electrodes 3 cm above the inion and 3 cm above and lateral. From Tyler et al. (Invest Ophthalmol Visual Sci 18:703, 1979) with permission of the authors and Lippincott/Harper and Row.*

similar to that obtained with steady-state VEPs to checkerboard reversal. The slope of the curve at high spatial frequencies allows extrapolation of a spatial frequency at zero amplitude which gives an electrophysiological measure of visual acuity that corresponds with behavioral acuity (Figure 10.7). The method can be used to study meridional, strabismic, and anisometropic amblyopia, ocular rivalry, fusion, and stereopsis.

---

## 10.5 VEPS TO BAR GRATINGS

Bar gratings consist of light and dark stripes of equal width with sharp borders between them. They produce VEPs less effectively than do checkerboard patterns because they present contrast borders in only one dimension. They are not used as often as are grating patterns with sinusoidal transitions because, although they represent mainly the spatial frequency represented by the grating itself, the sharp borders contain additional, higher, spatial frequencies that are probably processed by different channels of the visual system. Bar gratings are therefore much less specific for spatial frequency than are sinusoidal gratings.

Like VEPs to checkerboards, bar grating VEPs depend on the size of the stimulus element, the contrast between light and dark elements, the stimulus rate, and other parameters. However, VEPs to bar gratings and to checkerboard patterns have different sizes of effective stimulus elements. Study of maturation of visual acuity, as tested with VEPs to bar grating patterns, indicates that adult acuity is reached between four and six months of age, that is, at a time similar to that determined by checkerboard and sine wave grating patterns.

Clinical studies have used bar gratings only rarely. In ophthalmology, bar gratings have been used for refraction, for study of binocularity, and for demonstration that astigmatism reduces VEPs to stimulation with stripes in the direction of the lower visual acuity to an extent unexplained by the normal difference between the effects of oblique and horizontal or vertical stripes.

## 10.6 VEPS TO SMALL MACULAR LIGHT SPOTS

VEPs to stimulation with small light spots projected into the center of fixation have been compared with VEPs to check reversal and to diffuse light in patients with MS and retrobulbar neuritis. Macular stimuli were found to be more effective than diffuse light stimuli, but the outcome of comparisons with checkerboard patterns seems to depend on the size of the macular flash stimuli relative to that of the pattern element: Local flashes were inferior to checkerboard patterns when they were much larger than the checks used for comparison, equally effective when of the same size checks, and more effective when smaller than the checks; this suggests that the small light spot acts as a single element of a stimulus pattern. In one study, stimuli of less than 4° were found to be less useful clinically than larger ones because latencies became longer and more variable and fixation became a problem.

## 10.7 VEPS TO MOVING AND STEREOSCOPIC RANDOM DOT PATTERNS

VEPs may be induced by moving patterns. Especially effective are patterns of random dots that continue to appear and disappear and are aligned stereoscopically for the two eyes to give

the illusion of dots moving through space (dynamic random dot stereogram). The stimulus consists either of a sudden apparent displacement of the dots toward the observer (moving dot stereogram) or of the sudden alternation of the stereogram with a similar pattern without stereoscopic alignment of the images for each eye (dynamic random dot correlogram). This stimulus, sometimes called *cyclopean,* is so obtrusive that it does not require much effort of fixation by the subject and should be useful for the study of children and uncooperative adults; it has been used mainly for study of binocularity or stereopsis, a cortical function.

## 10.8 VEPS TO OTHER PATTERNED STIMULI

Printed letters and words have been used as stimuli for VEPs to distinguish between shape and perceptual content of the stimulus. Abnormal VEPs were found in reading-disabled children and adults. Similar stimuli have been used to study language and cognitive functions by recording late components such as the P300.

VEPs to stationary dots of various sizes were used to study visual acuity and refractive errors. Grid patterns of different sizes were presented to the two eyes to test stereoacuity. VEPs to horizontal lines were used to study cerebral asthenopia. Patterns of texture contrast elicit VEPs different from similar black-and-white patterns. VEPs to laser speckle patterns may become helpful in cases of cataracts and other opacities of ocular media.

## 10.9 PHOTOPIC VERSUS SCOTOPIC VEPS

The VEPs used in clinical practice are almost entirely due to the excitation of the photopic retinal cone system, located mainly in the fovea, which operates under conditions of light adaptation and is most sensitive to red. Dark adaptation, stimulation of the peripheral retina, dim stimuli, and blue color all favor excitation of the scotopic rod system, which contributes little to VEPs produced with conventional methods and has only rarely been tested in neurologic diseases such as MS.

## 10.10 COLOR VEPS

The effect of different wavelengths on transient and steady-state VEPs has been studied with diffuse and patterned stimuli. Colored stimuli have occasionally been used in the evaluation of neurological problems, especially of MS and in ophthalmological investigations of abnormal color vision.

## 10.11 VEPS TO BLINKING AND SHIFT OF GAZE

VEPs produced by interruption of visual input due to blinking differ from VEPs produced by interruption of input by other means.

Saccadic shifts of gaze, causing lambda waves in the occipital EEG of some subjects, produce VEPs that are partly, but not fully, explained by the change of the visual input. These VEPs are not identical to VEPs produced by other presentation modes of the same input, although the onset of the new input following the change of gaze, rather than the termination of the old input from the preceding fixation, appears to be the effective stimulus.

## 10.12 ELECTRIC STIMULATION OF THE EYE

Electric stimulation of the eye can be used to evoke occipital potentials.

## 10.13 THE ERG TO DIFFUSE AND PATTERNED LIGHT

The ERG may be elicited by diffuse flashes or by patterned light stimuli. Diffuse illumination of the entire retina with Ganzfeld stimuli produces larger VEPs than restricted, focal retinal stimulation. Recordings are made with electrodes on cornea, sclera, or near the eye. Care must be taken to avoid pickup of the ERG from the opposite eye.

The ERG to diffuse light flashes consists of a cornea-negative *a*-wave and a cornea-positive *b*-wave which are probably generated by photoreceptors and glial Muller cells, respectively. The scotopic rod system and the photopic cone system contribute to different degrees to these ERG waves. The photopic component can be separated from the scotopic component by stimulation with flashes of different wavelength under light-adapted and dark-adapted conditions. The ganglion cells of the optic nerve do not participate in the production of the flash ERG. The ERG may be recorded simultaneously

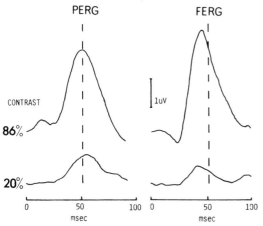

FIGURE 10.8. *Electroretinograms to patterned and diffuse light. The pattern ERG (PERG) and focal ERG to diffuse light (FERG) were produced by presenting a checkerboard pattern or a diffuse light flash in a round 16° field. Both ERGs show mainly a b-wave. This is preceded by a small a-wave of opposite polarity, visible at the higher contrast level, mainly in the FERG. Contrast levels refer to simultaneous brightness contrast between light and dark phase for the FERG. The tracings are averages of recordings from three normal subjects. Recording between a gold foil electrode on the sclera and an electrode on the eyebrow. Positivity at the sclera is plotted upward. From Arden et al. (Ann NY Acad Sci 388:580, 1982) with permission of the authors and New York Academy of Sciences.*

with the VEP and may aid in the distinction between ocular and central visual problems.

The ERG to patterned light differs from that to diffuse light (Figure 10.8) and varies with the size of the pattern elements. The effective size of the pattern elements for the ERG differs from that for the VEP. In contrast to the flash ERG, the pattern ERG is probably partly or entirely produced by ganglion cells or some preganglionic elements. ERGs to pattern stimulation have been found to be abnormal in amblyopia, optic neuritis, some cases of MS, glaucoma, and various other disorders involving the optic nerve in or near the eye. Because the pattern ERG indicates the entry of the afferent volley into the central visual pathway, it may be used to calculate retinocortical conduction time by subtracting the latency of the pattern ERG from the latency of the simultaneously recorded pattern VEP. This central conduction time may help to distinguish retinal lesions, such as maculopathy, which increase both retinal and cortical latency, from central lesions, such as optic nerve demyelination, which increase only the cortical latency.

# C

# Auditory Evoked Potentials

---

## PART CONTENTS

# AEP Types, Principles, and General Methods of Stimulating and Recording

## 11.1 AEP TYPES

The most common AEP types are listed in Table 11.1. AEPs are distinguished mainly by recording electrode placement and latency. Recordings between electrodes on the vertex and the earlobe or mastoid process yield AEPs of three latency and amplitude ranges (Figure 11.1). The short-latency AEPs include peaks of up to 10 msec and of about 0.2 μV; they are generated in the brainstem. The middle-latency AEPs have several variable peaks with latencies of 10–50 msec and with amplitudes of about 1 μV; they probably reflect early cortical excitation. The long-latency AEPs, beginning after 50 msec and having peaks of 1–10 μV, represent later cortical excitation. These three kinds of potentials are sometimes referred to as early, middle, and late AEPs. The ECochG and the sonomotor AEPs are recorded with electrodes placed near the cochlea and the neck or scalp muscles, respectively. In contrast to VEPs, AEPs are classified by recording methods, not by stimulus characteristics: Most AEPs are elicited with a click stimulus.

TABLE 11.1. AEP types

A. Short-latency (up to 10 msec) AEPs
  1. Transient short-latency AEPs
    a. Brainstem AEP (BAEP)
    b. The slow brainstem AEP
  2. Steady-state short-latency AEP: Frequency following potential (FFP)
B. Middle-latency (10–50 msec) AEPs
  1. Transient middle-latency AEP (MLAEP)
  2. Steady-state middle-latency AEP: 40 Hz AEP
C. Long-latency (over 50 msec) AEPs
  1. Transient long-latency AEP at 100–200 msec (LLAEP)
  2. Transient AEPs of longer latency
D. Electrocochleogram (ECochG)
  1. Acoustic nerve action potential (NAP)
  2. Cochlear microphonic
  3. Summating potential
E. Sonomotor AEPs
  1. Postauricular AEP
  2. Neck and scalp sonomotor AEP

## 11.2 THE SUBJECT

During the recording of the AEP, the subject either reclines in a comfortable chair or lies on

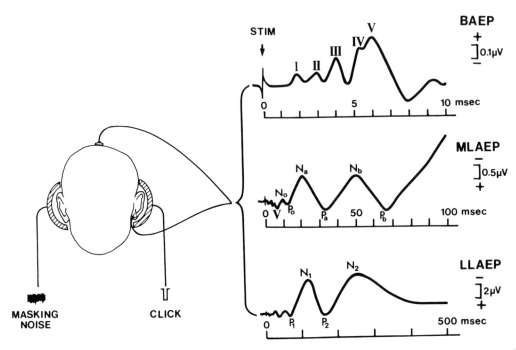

FIGURE 11.1.   *Schematic diagram of the BAEP, MLAEP, and LLAEP. Stimulation of one ear and recording between the ipsilateral earlobe and the vertex produces sequences of several groups of peaks that can be displayed at three recording speeds and gain settings. Recording with a short time base at a high gain shows the first five peaks (I–V) that occur in the first 10 msec and represent the BAEP. Recording with a medium time base at a slightly lower gain shows the middle-latency AEP (MLAEP), which occurs between 10 and 50 msec after the stimulus. Recording with a long time base at an even lower gain shows the long-latency AEP (LLAEP) beginning more than 50 msec after the stimulus. Positivity at the vertex is plotted upward for the BAEP. Note that the MLAEP and LLAEP are displayed with a polarity opposite to that used for the BAEP. The MLAEP and LLAEP are preceded by the small peaks of the earlier AEPs. The diagrams are characteristic for stimulation with rarefaction clicks at about 60 dB SL and continuous contralateral masking noise.*

a bed in a quiet room. In most recordings, special care must be taken to relax neck muscles by placing pillows under the head and adjusting the position of the body. The subject wears earphones or, rarely, listens to a loudspeaker. The recording room should be so quiet that the subject can not hear any sounds except the stimulus and the masking noise. Earphones must be applied cautiously in children to avoid collapse of the outer ear canal. Because sleep relaxes scalp muscles and reduces biological artifacts without altering the BAEP and the ECochG, subjects are encouraged to sleep during these recordings. Sleep is often induced with common sedatives, especially in infants and young children. The MLAEP may be recorded in light

sleep, but its threshold increases in deep sleep. LLAEPs vary not only with sleep but even with changes of attention. Sonomotor AEPs must be recorded while the subject is awake and not relaxed.

## 11.3 STIMULATION METHODS

### 11.3.1 Stimulators

#### 11.3.1.1 Earphones

The stimulus is most commonly delivered through an earphone. Earphones are usually of the moving-coil type, which has low impedance and, especially at high stimulus intensities, generates electromagnetic fields that induce stimulus artifacts that may require shielding of the earphone with mu metal. Electrostatic and piezoelectric earphones, having high impedance, require less current but higher signal voltage, and may not be able to generate sound levels as high as those produced by an electrodynamic earphone. These earphones produce mainly electrostatic stimulus artifacts that are more easily eliminated by shielding. In some laboratories, the stimulus artifact is reduced by separating the source of the sound from the ear by a piece of tubing that introduces a delay between the electric production of the sound and the AEP. Small earphones that fit into the external ear canal may be used for intraoperative monitoring of BAEPs.

### 11.3.1.2 Loudspeakers

Loudspeakers are used only rarely for acoustic stimulation. This stimulating method does not allow testing of monaural AEPs except in patients with unilateral deafness and in recordings of the ECochG. Furthermore, this stimulating method requires correction of stimulus intensity and response latency for the distance between the stimulus source and the ear.

### 11.3.1.3 Bone vibrators

Stimulators transmitting sound waves through a pressure foot to the mastoid bone have been used to evaluate bone conduction and to estimate conductive hearing loss. However, the stimulus intensity and frequency reaching the cochlea are difficult to control. Bone stimulation is therefore not used routinely but may be helpful when air stimulation can not be used, for instance, during operations and in cases of malformations of the external ear.

### 11.3.1.4 Calibration

The sound stimulus intensity at the tympanic membrane depends on the acoustic coupling between the sound stimulus generator and the ear. Audiometric zero intensities for air and bone conduction are standardized with specific cou-

plers that roughly resemble the human ear and mastoid (artificial ear). The stimulus intensity measured with this method refers to the sound pressure generated by the various earphones and vibrators.

### 11.3.2 Stimulus Types

Several types of stimuli may be used to elicit AEPs (Figure 11.2). Most AEPs are produced by clicks or tone pips. Click stimuli are very satisfactory for neurological studies because they produce sudden excitation resulting in a well-defined EP. However, clicks are not very suitable for audiological studies because they contain a wide range of tone frequencies, act mainly by virtue of their high-frequency content, and do not test the lower frequency range, which is important for speech. Stimulation with tones of lower frequency, although desirable for audiological purposes, creates several problems. Tones act as a stimulus mainly by virtue of their onset and should be short. However, expression of the tonal frequency of a stimulus requires a tone duration of at least a few cycles. Tones of low frequency have a wavelength too long for an effective stimulus, especially in cases of short-latency AEPs. Furthermore, although a stimulus of sudden onset is needed for a clearly defined EP, the sudden onset of a loud tone of any frequency introduces a high-frequency transient and thereby elicits an AEP that is not specific for the tone frequency. A compromise between the requirements of a specific tonal frequency and a well-defined onset is available in the form of the filtered click, the tone pip, and the logon, although low frequencies remain a poor stimulus for excitation of AEPs. The receptors for low tones, located in the apical part of the cochlea, are not easily explored by AEP methods. Because of the longer travel of sound waves to the apex of the cochlea, responses to low tones are elicited later than responses to high frequencies, which excite the basal part of the cochlea. Also, because low-frequency sound waves have a long duration, the responses are dispersed in time and have a low amplitude. Therefore response components to the low-frequency contents of an auditory stimulus are obscured easily by the earlier and larger response components generated by the more basal parts of the cochlea responding to the high-frequency content represented by the sudden onset of the stimulus. Attempts have been made to study BAEPs to low stimulus frequencies by masking high-frequency components of the stimulus with continuous noise or to record slow brainstem AEPs to low-frequency tone pips.

### 11.3.2.1 Broadband clicks

Clicks may be produced by feeding electric monophasic square pulses into an earphone and thereby deflecting the earphone membrane. Even though the electric pulses are sharp and last only 100 μsec, the earphone membrane reacts to the square pulse only imperfectly and generates pressure changes that are further modified by the material used for coupling the earphone to the head and by the intervening ear structures before they reach the sound receptors in the cochlea. These factors convert the original electric square pulse into a sequence of decaying pressure fluctuations with a peak acoustic power at 2–4 kHz (Figure 11.2*a*). This sound stimulus still contains a fairly wide range of frequencies and is therefore called *broadband* or *unfiltered click*. A narrower band of stimulus frequencies can be obtained either by filtering the electric square pulse or by presenting broadband clicks on a background of a continuous masking noise that eliminates the effect of some of the frequencies in the click.

### 11.3.2.2 Filtered clicks and tone pips

Filtered clicks (Figure 11.2*b*) are generated by simply passing a rectangular or sinusoidal wave through a filter with a narrow bandpass so as to produce a brief burst of waves of a frequency centered at the filter bandpass. Tone pips (Figure 11.2*c*) have more symmetrical rising and falling phases and are produced either by passing one period of a sine wave through a bandpass filter of the same frequency or by modulating the amplitude of a pure tone electronically to give it the desired rise, fall, and plateau times. Filtered clicks and tone pips are usually given rising and falling phases of two cycles and a plateau of one cycle. A logon (Figure 11.2*d*) is a tone with an amplitude modulated by the shape of a Gaussian distribution curve, said to give the best compromise between the definitions of stimulus onset and frequency. Filtered clicks, tone pips, and logons must always start from the zero level and take off in the same direction to give consistent AEPs.

### 11.3.2.3 Tone bursts

Tone bursts (Figure 11.2*e*) with rise and fall times of at least 5 msec and durations of at least 30 msec may be used for the LLAEP but are not suited for the MLAEP and BAEP.

### 11.3.3 Stimulus Rate

The stimulus frequency varies for each AEP type. Transient short- and medium-latency AEPs

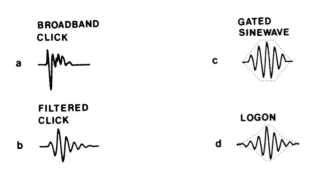

**BROADBAND CLICK**

a

**FILTERED CLICK**

b

**GATED SINEWAVE**

c

**LOGON**

d

FIGURE 11.2.   *Acoustic waveforms of auditory stimuli.* (a) *Broadband click;* (b) *filtered click at 2 kHz;* (c) *electronically gated sine wave at 2 kHz with a rising phase of 2 cycles, a plateau of 1 cycle, and a falling phase of 2 cycles;* (d) *logon at 2 kHz with an envelope of a probability curve;* (e) *tone burst of 0.5 kHz with rising and falling phases of 2 cycles.*

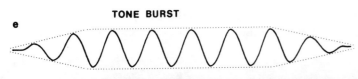

**TONE BURST**

e

such as the BAEP, slow brainstem AEP, ECochG, MLAEP, and sonomotor AEP are usually elicited at rates of 8–10/sec. The LLAEP requires stimulation at 1/sec or less. The steady-state 40-Hz AEP is produced by tone bursts of various frequencies repeated at 40/sec; the FFP can be obtained with tones of 100–1,000 Hz. In the choice of the exact stimulus rate, one should avoid synchronization with power-line frequency because it could lead to buildup of interference artifact in the average. Although the stimulus frequency is usually not altered during a stimulation, a brief tone of changing frequency and a change of the frequency of a sustained tone are also capable of eliciting AEPs.

## 11.3.4 Stimulus Intensity

Although the stimulus intensity is usually not changed during the recording of an AEP, a change of the intensity of a continuous tone may also act as a stimulus and elicit an AEP.

Stimulus intensity is measured as a ratio between stimulus level and a reference level and is expressed in decibels. The number of decibels equals 20 times the logarithm$_{10}$ of this ratio when expressing amplitude, or 10 times the logarithm$_{10}$ of this ratio when expressing power or the square of voltage. For instance, a stimulus with an amplitude 1,000 times the amplitude of a reference level is 60 dB above that level. The same type of measurement is used to indicate hearing loss: A subject hearing a tone only if it is increased by 40 dB above the normal hearing level is said to have a 40-dB hearing loss for that tone. Several reference levels are used to describe stimulus intensity in AEP studies.

### 11.3.4.1 Hearing level

The hearing level (HL), routinely used in audiometry, is the average threshold intensity of normally hearing young adults tested with pure tones of at least 0.5 sec.

### 11.3.4.2 Normal hearing level

The normal hearing level (nHL) for pure tones of 0.5 sec or more is the same as the HL. However, the normal hearing level for shorter

tones and for clicks and other less regular sounds used as stimuli for AEPs differs from that for pure tones of the same amplitude. Therefore, the normal hearing level for these stimuli must be defined by measuring the average threshold intensity for the specific stimulus in a group of audiologically normal young adults. If nHL is to be used as a reference for EP studies, the manufacturer or the user of the stimulating equipment must make such measurements to calibrate the equipment and give a valid intensity reference.

### 11.3.4.3 Sensory level

The sensory level (SL) is the hearing threshold of an individual. In a normally hearing young subject, the SL for sustained pure tones is the same as the HL, and the SL for other stimuli is the same as the nHL. In a subject with reduced hearing, the difference between SL and HL or nHL expresses the magnitude of the hearing loss for the particular stimulus. Hearing threshold is measured by increasing and decreasing stimulus intensity in 5-dB steps and determining the midpoint of the intensities at which the subject begins or ceases to hear the stimulus.

Because SL cannot be measured in infants and young children, one of the other references must be used for stimulus intensity settings. In adults, the use of SL as a reference of stimulus intensity also has some problems.

### 11.3.4.4 Sound pressure level

The sound pressure level (SPL) defines the intensity of the sound stimulus physically but does not relate directly to the physiological effect of the stimulus. An arbitrary zero reference point is conventionally set at 0.00002 dynes/cm$^2$ (20 microPascals) for a tone of 1,000 Hz. The normal hearing threshold exceeds this level by a number of decibels that varies with the frequency of sustained sound stimuli such as pure tones.

### 11.3.4.5 Peak equivalent sound pressure level

The pressure of brief sounds is difficult to measure exactly because sound level meters have a minimum response time. One method to overcome this problem is to compare the peak-to-peak amplitude of the stimulus with the amplitude of a sine wave of a pure tone having the same peak-to-peak amplitude as the click. Measurements of peak equivalent sound pressure level (peSPL) do not relate closely to the nHL but usually give higher hearing thresholds because shorter stimuli must have higher amplitude to reach hearing threshold.

### 11.3.5 Stimulus Polarity: Condensation and Rarefaction Clicks

Click stimuli may be produced by electric pulses, which cause an initial deflection of the earphone membrane toward the eardrum, condensing or compressing the air in the ear canal and generating condensation, or compression, clicks. If the polarity of the electric pulse driving the earphone membrane is reversed, the pulse produces an initial deflection of the membrane away from the ear, rarefying the air in the ear canal and causing rarefaction clicks. The manufacturer or user must identify the electric polarity settings that produce the rarefaction or condensation clicks. Stimulus polarity is practically important for short-latency AEPs because condensation and rarefaction clicks produce slightly different BAEPs and ECochGs.

### 11.3.6 Masking Noise

When a sound stimulus is applied through an earphone to one ear, the sound is conducted by the skull and may reach the opposite ear. Although the stimulus is attenuated by about 50–60 dB on its travel across the head, it may excite the other ear. Such cross-stimulation is especially likely to occur when the ear to be tested has a higher threshold than the other ear and is exposed to strong stimuli. Cross-stimulation can be avoided by applying a constant masking noise through an earphone on the ear opposite the stimulated one. The masking noise should have an intensity of about 40 dB below the stimulus intensity. Masking intensities of about 60-dB SPL are sufficient for the stimulus intensities used in routine studies. Masking noise should contain either a wide range of frequencies at equal intensities (white noise) or include at least the frequencies of the stimulating sound. Masking is not necessary for the ECochG because the response is generated and recorded only at the stimulated ear.

Masking noise may be used to mask unwanted frequencies of a complex sound stimulus by mixing masking noise and stimulus sound at the same ear.

### 11.3.7 Monaural and Binaural Stimulation

Separate stimulation of each ear is needed for studies evaluating hearing, acoustic nerve, and brainstem function on each side. Simultaneous stimulation of both ears produces AEPs that differ from the sum of the AEPs to stimulation of each ear.

## 11.4 RECORDING ELECTRODE PLACEMENTS, MONTAGES, AND POLARITY CONVENTION

For recordings of most AEPs, electrodes are placed on the vertex and on or near the ears. Ear electrodes may be placed on the lateral or medial surface of the left and right earlobe (A1, A2) or on the scalp over the left and right mastoid bone (M1, M2). Placement on the medial surface of the earlobe can reduce the stimulus artifact of the click. If necessary, the definition of the first wave of the BAEP may be improved by recording the ECochG from an electrode inserted into the ear canal or through the tympanic membrane. Sonomotor AEPs are recorded with electrodes placed over neck or scalp muscles.

The same electrode combinations are used to record BAEPs, MLAEPs, and LLAEPs. For single channel recordings, the vertex electrode and the electrode at the stimulated ear are connected to the two inputs of the amplifier and the electrode at the opposite ear is connected to the amplifier ground. Dual-channel recordings are preferred for BAEPs, channel 1 recording between vertex and stimulated ear, channel 2 between vertex and nonstimulated ear. A ground electrode is placed anywhere on the head or elsewhere on the body and is connected to the amplifier ground.

The gain and filter settings, sweep length, and number of responses averaged vary with the AEP type. For every recording, at least two sets of averages must be superimposed to ascertain replication.

American EEG Society guidelines do not recommend one specific polarity convention, however most laboratories record the BAEP so that an upward deflection indicates increased positivity at the vertex electrode. This makes the relevant peaks convex upward. MLAEPs and LLAEPs are usually recorded with the opposite polarity convention: Upward deflections indicate increased vertex negativity (Figure 11.1).

# 12

# The Normal BAEP

## 12.1 DESCRIPTION OF THE NORMAL BAEP

The most commonly used methods of studying the BAEP are summarized in Table 12.1. Examples of normal BAEPs are shown in Figure 12.1. The BAEP recorded on the side of the stimulated ear contains five wavelets appearing in the first 10 msec after the stimulus and having peaks that are positive at the vertex with reference to the ear. These waves are usually labeled with roman numerals I to V. The BAEP recorded from the side opposite the stimulated ear differs from the ipsilateral BAEP mainly in that it shows no clear-cut wave I; the negative peak preceding wave II ($I_N$) is therefore sometimes used for measurements of interpeak latencies. Other peaks differ slightly but significantly on the two sides.

Not all normal recordings contain all BAEP peaks. Wave V is present most often, waves I and III can usually also be identified. Wave II is often absent and wave IV may merge more or less completely with wave V. Wave V is sometimes followed by waves VI and VII. Peak and interpeak latencies of the BAEP waves are remarkably constant for a given set of subject, stimulus, and recording parameters. They vary little in repeated recordings in the same subject even over many months; they vary only slightly more between recordings from the two sides of the same subject and among different subjects of the same age and sex. Latency is longer and more variable for condensation than for rarefaction clicks. Each laboratory must therefore establish its own normal values for these variables.

Because not all BAEP waves are present in every recording from a normal subject, a few practical maneuvers are often helpful in enhancing and identifying BAEP waves:

- Wave I may be enhanced by increasing the stimulus intensity and decreasing the stimulus rate. Recording a BAEP to condensation clicks in addition to the BAEP to rarefaction clicks may help to distinguish wave I from mechanical and electric stimulus artifacts and from cochlear microphonic: Artifacts and cochlear microphonic reverse polarity with

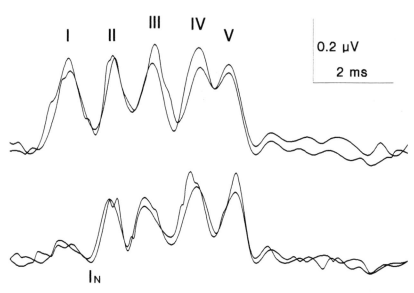

FIGURE 12.1. *Normal BAEP to stimulation of the right ear, recorded between vertex (Cz) and right ear (A2) (top tracing) and vertex and left ear (A1) (bottom tracing). The contralateral recording shows absence of wave I, presence of a negative wave preceding wave II ($I_N$), earlier appearance of wave III, and better separation of waves IV and V. Click stimuli of 100 μsec duration and 60 dB SL with 40 dB SL continuous masking noise to the opposite ear. Each tracing is an average of 2000 responses. Two tracings obtained from successive stimulation are superimposed. Recording bandwidth is 100–3000 Hz. Relative positivity at the vertex gives an upward deflection. This subject is a normal 33-year-old woman.*

reversal of stimulus polarity whereas wave I does not; however, wave I may change latency with stimulus polarity reversal, especially in subjects with hearing loss. Averaging responses to alternate rarefaction and condensation stimuli, although not recommended as a routine procedure, may be the only way to resolve wave I in BAEPs to strong stimuli which tend to produce longer stimuli artifacts and to shorten the latency of wave I. Comparisons with recordings between vertex and nonstimulated ear, which do not contain wave I, may help to identify wave I in recordings between vertex and stimulated ear. Recording between the two ears augments wave I relative to other waves. In exceptional cases, simultaneous recording of the ECochG may help to identify wave I.

- Wave II, although often absent in normal subjects, may be of clinical significance if it shows prolonged absolute latency of IPL I–II.

Wave II is enhanced in recordings between vertex and contralateral ear that do not show wave I.

- Wave III may be normally split into two peaks; its latency is then measured to the first peak or to the middle between the two peaks. Splitting may disappear if condensation clicks are used instead of rarefaction clicks and vice versa. A split, or bifid, wave III must be distinguished from a partial fusion of waves III and IV or of waves II and III. Fusion of waves II and III occurs especially in recordings between vertex and nonstimulated ear.
- Wave IV often normally fuses with wave V to form a complex with a peak latency equivalent to that of wave IV, wave V, or an intermediate value. Recording between the vertex and the nonstimulated ear (Figure 12.1), or stimulating with clicks of the opposite phase, may produce better separation of waves IV and V. Wave IV often varies in the same individual with time.

- Wave V is the most reliable peak. It may be identified by its low threshold, its persistence during repetitive stimulation up to 100/sec, and by the large negativity that commonly follows it. Occasionally, wave V consists of only a small inflection on the downslope of wave IV. Recordings from the nonstimulated ear may give a clearer definition of wave V. An unusually large negativity following wave IV may be a normal variant and may disappear in recordings from the nonstimulated ear or on stimulation with clicks of the opposite polarity. The variation of wave V with click polarity is greater in patients with abnormal BAEPs than in normal subjects.

## 12.2 SUBJECT VARIABLES

### 12.2.1 Age

#### 12.2.1.1 Before adulthood

BAEPs change considerably with age.[85] The rate of change, and the variability among subjects of the same age, are largest early in life and decrease with age. BAEPs may be absent in normal premature infants under 30 weeks of conceptional age. BAEP peak and interpeak latencies decrease steadily between conceptional ages of 25 and 44 weeks; at each age, the latency decreases to a different degree with increasing stimulus intensity.[24,66] The IPL I–V shortens by 2–3 msec over this period, and by about 1 msec during the last 6 weeks before term; the weekly decrease of latency averages 0.45 msec around 32–34 weeks but less than 0.1 msec near term. In full-term newborns, IPL I–V is about 0.8–1 msec longer than in adults, due mainly to a longer latency of wave V. Latencies continue to decrease after term. The latency of different waves decreases at different rates, in a manner indicating that the peripheral and central parts of the auditory path mature at different speeds. The latency of wave I, or peripheral conduction, reaches adult level at about 6 weeks of age, whereas the IPL I–V, or central conduction, reaches that level at about 1.5 years or somewhat later. The increase of latency with fast repetitive stimulation is also greater at birth than in adults. Age-specific normative values should therefore be established for intervals of 1 week in the preterm period, and for the ages of 3 weeks, 6 weeks, 3 months, 6 months, and 1 year thereafter. Maturation of the BAEP is not affected by prematurity, the maturity depending on conceptional age.[31]

#### 12.2.1.2 Aging

Advancing age has often been reported to increase both peak and interpeak latencies,[72] however, a relationship between latencies and advancing age is not universally accepted. In general, after early development, age has little effect on peak latencies and virtually no effect on interpeak latencies. Amplitude is reported to decrease with age. Because of these conflicting reports, no general rules for age-specific standards have been established. Age-corrected norms for adults are not in routine use.

### 12.2.2 Sex

Females, beginning in infancy or as late as about 8 years old, usually show BAEPs of shorter peak and interpeak latency and of higher amplitude than males, possibly due to sex-related differences in skull and brain anatomy. Different sets of normative data are used to evaluate BAEPs of males and females in laboratories that find that this difference is significant and increases the diagnostic power of BAEPs.

### 12.2.3 Body Temperature

A decrease of body temperature by 1° may increase the latency of wave V by 160 µsec or by 200 µsec and the IPL by 160 µsec. Variations of body temperature may explain circadian variations of these latencies.[54]

### 12.2.4 Hearing

Patients with hearing loss should have an audiological examination to investigate the peripheral or central causes. AEPs may help in this investigation. Peripheral hearing losses tend to increase BAEP peak latencies rather than IPLs and therefore do not necessarily preclude the identification of central lesions. However, significant hearing losses may make a reliable interpretation of BAEPs impossible.

Hearing loss may result in a dispersion of the action potential volley in the acoustic nerve, at times producing separated volleys. This may result in an altered waveform, often with peaks that are difficult to identify precisely.

## 12.3 STIMULUS CHARACTERISTICS

### 12.3.1 Effect of Stimulus Type

Broadband clicks, produced by delivering an electric square pulse of 100 μsec into an audiometric speaker, are the type of stimulus most commonly used in neurological and audiological studies. These clicks act mainly by virtue of their high-frequency content. Filtered clicks or tone pips are sometimes used for audiologic studies in an attempt to improve the frequency specificity of BAEPs to stimuli of tonal frequencies of 500–2,000 Hz. However, tones in this relatively low frequency range produce BAEPs that are due mainly to the high-frequency components produced by the sudden onset of a loud tone. Attempts have been made to study the effects of lower stimulus frequencies by presenting broadband clicks on a background of a continuous masking noise of high frequencies which eliminate the effects of unwanted high frequencies contained in the click stimulus. By subtracting the AEP to stimuli delivered with a masking noise of a specific frequency from the AEP to the same stimulus delivered with a masking noise of higher frequency, one may derive an AEP representing a narrow range of stimulus frequencies. A more direct method than this technique of "derived BAEPs" attempts to obtain frequency-specific BAEPs by mixing the stimulus with masking noise filtered with a notch filter that eliminates a narrow band of frequencies. A broadband transient stimulus given on the background of this continuous masking sound acts by virtue of that narrow part of the auditory spectrum that is not rendered ineffective by the masking sound.

### 12.3.2 Effect of Stimulus Rate

The stimulus rate is usually set at 8–10/sec, avoiding fractions of 60/sec that could lead to synchronization with interference from power lines and build up 60-Hz artifact in the average. An increase of stimulus rate above 30/sec increases the latency and decreases the amplitude of the BAEP. The increase of latency is best seen in wave V because earlier waves may be difficult to distinguish at these high rates. However, some studies have demonstrated that earlier waves show different degrees of rate-dependent latency shifts, causing an increase of the IPL I–V at high stimulus rates and suggesting that peripheral and central conduction undergo different degrees of adaptation.

Low stimulus rates of about 4/sec may improve the definition of BAEPs in some instances.

### 12.3.3 Effect of Stimulus Polarity

BAEPs to rarefaction clicks are used more often because they have shorter latency and clearer definition than BAEPs to condensation clicks.[83] The method of averaging responses to clicks of alternating phase, often used to reduce the stimulus artifact and cochlear microphonic, is not generally used as a routine because it leads to the addition of two kinds of responses and may result in partial cancellation and superimposition of peaks of different latency and shape. Click polarity does not affect threshold.[78] In a few patients, abnormal BAEPs appear only with one stimulus polarity but not with the other.

### 12.3.4 Effect of Stimulus Intensity

Stimulus intensity affects both amplitude and latency. Wave V has the lowest threshold: It appears at, or slightly above, the hearing threshold; the other waves appear at higher intensities and reach maximum amplitude at different levels, usually at medium intensity. Wave IV may grow with increasing intensity after wave V has reached maximum; if it merges with wave V, it may appear to shift the latency of wave V to an earlier value. The amplitude of wave I continues to grow as stimulus intensity grows to above 60 dB SL. In patients with hearing deficits, high stimulus intensities may be needed to elicit a BAEP.

The latency of all BAEP waves decreases with increasing stimulus intensity up to about 60 dB SL. The latency of wave V lies slightly below 6 msec in normal subjects past infancy. The latency decrease is similar for waves II to V so that IPLs between these waves are generally fairly independent of stimulus intensity. However, the latency of wave I increases rather abruptly, and more than that of other waves, at stimulus intensities below 60 dB SL so that IPLs between wave I and other waves diminish with lower stimulus intensities, especially when stimuli of low tonal frequency are used. Latency and amplitude do not correspond with the subject's estimate of stimulus loudness.

### 12.3.5 Effect of Monaural and Binaural Stimulation

The BAEP to binaural stimulation has a distribution different from that of BAEPs to monaural stimulation[69] and differs from the sum of BAEPs to monaural stimuli mainly in the later BAEP waves. This binaural interaction may represent activation of brainstem connections not excited by stimulation of either ear. Because binaural stimulation only rarely provides clinically useful information, and may mask monaurally detected abnormalities, it is not used routinely in most laboratories.

## 12.4 RECORDING PARAMETERS

### 12.4.1 Electrode Placements: Ipsilateral and Contralateral Recordings

BAEPs to stimulation of each ear are recorded routinely from both sides of the head with the same electrode placements as used for most other AEPs (Table 12.1). Recordings between the ear electrodes may be added to better define wave I.

Recordings from the two sides differ in several regards and seem to be generated mainly by auditory structures ipsilateral to the stimulated ear. In addition to the unilateral appearance of wave I at the electrode near the stimulated ear,

peak latencies of waves III and IV are shorter and peak latencies of waves II and V are longer in contralateral recordings, causing shortening of IPL II–III and lengthening of IPL IV–V. This difference is probably not entirely explained by differences in the conducting media on the two sides but is probably due to activity of lateralized structures in the brainstem.

Topographic studies have shown that the BAEP waves following wave I have a wide distribution. Unlike potentials generated in the cortex near one electrode, these far-field potentials can not be attributed to either the ear electrode or the vertex electrode.

### 12.4.2 Filter Settings

The bandpass should include frequencies between 10–30 Hz and 2,500–3,000 Hz avoiding steep filter roll-offs. The low frequency cutoff may be raised to 100–200 Hz if otherwise irreducible muscle or mechanical artifact obscures the BAEP. Narrowing filter settings may distort amplitude and latency of BAEP peaks.[17] A low-frequency filter setting of 100 Hz usually causes no more than minimal distortions of latency, but may be too high for slow components, especially those in BAEPs to low-frequency tone pips. Although filters affect absolute peak latencies more than IPLs because they shift latencies of similar peaks to a similar degree[16] they can not help but also change the IPLs because the peaks are not identical in configuration and therefore are not shifted by exactly the same amount.

It seems surprising that low-frequency filter settings of only 200 Hz and less should affect the BAEP which consists of waves that have a duration of only 1–2 msec corresponding with wave frequencies of no less than 500 Hz. However, these faster waves are mixed with less obvious slow-wave components which are important for the configuration and the timing of the faster waves. The considerable contribution by slow components to the BAEP is clearly revealed by power spectral analysis of the BAEP.[13]

### 12.4.3 Other Recording Parameters

Two channels should be used to record between vertex and stimulated ear and between vertex

TABLE 12.1.  BAEPs to clicks

---

A. Subject variables
   1. Age: Several separate normal control groups should be used for infants under 18 months and for elderly subjects as needed in each laboratory.
   2. Sex: Separate controls for males and females are used in some laboratories.
   3. Hearing: Hearing loss should be investigated audiologically to help distinguish between peripheral and central causes.
B. Stimulus characteristics
   1. Type: Broadband clicks produced by 100 μsec electric square wave pulses into an audiometric ear speaker
   2. Rate: 8–10/sec
   3. Polarity: Rarefaction clicks; alternating clicks only if stimulus artifact to either polarity is intolerable
   4. Intensity: 115–120 dB peSPL or 60–70 dB SL; additional intensities for threshold measurements and latency-intensity curves in audiologic investigations
   5. Masking: 60 dB SPL of white noise to the contralateral ear
   6. Monaural stimulation: Monaural stimuli should be used routinely.
C. Recording parameters
   1. Number of channels: 2
   2. Electrode placements
      a. Left and right earlobes (A1, A2) or mastoid processes (M1, M2)
      b. Vertex (Cz)
      c. Ground electrode at Fz or elsewhere on the head or body
   3. Montages
      Channel 1: Cz–ipsilateral ear or mastoid process
      Channel 2: Cz–contralateral ear or mastoid process
   4. Filter settings
      a. Low-frequency filter: 10–30 Hz; 100–200 Hz in case of irreducible EMG or mechanical artifact
      b. High-frequency filter: 2,500–3,000 Hz
   5. Number of responses averaged: 2,000 (1,000–4,000)
   6. Sweep length: At least 15 msec
   7. Sampling rate: Minimum sampling rate is 10 kHz; maximum sampling interval is 100 μsec.
D. Analysis
   1. Normal peaks: I, III, and V in ipsilateral recordings; $I_N$, III, and V in contralateral recordings
   2. Criteria for abnormal BAEPs
      a. Central lesions
         (1) Absence of waves I–V, unexplained by extreme hearing loss or technical problems
         (2) Absence of waves following wave I or wave III
         (3) Abnormally prolonged IPL I–V
         (4) Abnormal decrease of the V/I amplitude ratio, especially when associated with other abnormalities
         (5) Abnormally long interaural difference of IPL I–V unexplained by middle- or inner-ear problems identifiable by audiometric tests
         (6) Questionable criteria: Abnormally long peak latency V on rapid repetitive stimulation; abnormally long peak latency III and V; abnormally increased IPLs I–III, III–V
      b. Peripheral lesions
         (1) Increased BAEP threshold
         (2) Increased latency of waves I–V
         (3) Abnormal latency-intensity curves

and nonstimulated ear. About 1,000–4,000 responses must be averaged. The sweep length should be at least 15 msec with a maximum dwell time of 100 μsec per point or a minimum sampling rate of 10 kHz, which can resolve waves up to 5 kHz; even an averager with only 250 points per sweep satisfies these conditions.

## 12.5 GENERAL STRATEGY

The BAEP is better suited than other AEPs for the detection of lesions in the lower parts of the auditory pathway. It is also the best AEP currently available for electric response audiometry. Although the BAEP is generated in the brainstem, a strict correlation between each wave and a specific brainstem structure in the auditory pathway is not possible. Earlier studies had attempted to relate wave I to the acoustic nerve, wave II to the cochlear nucleus and trapezoid body, wave III to the superior olive, wave IV to the lateral lemniscus, wave V to the inferior colliculus, wave VI to the medial geniculate body of the thalamus, and wave VII to the medial geniculate or auditory radiations. However, later studies have raised much doubt about the simple wave-to-

point correlations. Even the relationship between wave I and the acoustic nerve is not very clear. There is evidence to suggest that wave II is at least partly generated by the intracranial portion of the acoustic nerve and that wave I may be preceded by an earlier wave that may also be due to excitation of the acoustic nerve. Direct recordings from the vicinity of presumed generators in humans have shown that it is difficult to make unequivocal correlations between individual waves and brainstem structures. It is now believed that most waves are probably generated by more than one anatomical structure and that each structure may contribute to more than one wave. For clinical purposes, waves I, III, and V have been used to localize lesions to the area between acoustic nerve and upper brainstem and, in some cases, to suggest localization within the upper and lower parts of that area (Figure 12.2). However, these correlations between waves and level of a lesion within the brainstem remain tentative.

The side of a lesion is best determined by recording BAEPs to stimulation of each ear because the BAEP reflects excitation mainly of the brainstem auditory pathways on the stimulated side.

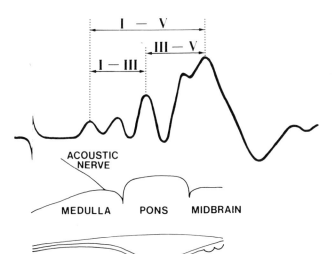

FIGURE 12.2. *Schematic diagram of the strategy attempting to localize lesions with the BAEP. A click stimulus to the ear elicits a sequence of five waves generated by the acoustic nerve and brainstem structures. IPL I–V represents conduction from the acoustic nerve to the upper midbrain; IPLs I–III and III–V represent conduction through the lower and upper brainstem, respectively.*

### 12.5.1 The Acoustic Nerve and Cochlear Nuclei

Severe lesions of the acoustic nerve may abolish the BAEP entirely or leave only wave I intact. If necessary, the action potential of the acoustic nerve can be searched for with special procedures. Less severe lesions may increase the latency of BAEP waves, starting either with wave I or with wave II. If the delay starts with wave I, IPLs are unchanged; if it starts with wave II, IPLs I–III and I–V are increased.

### 12.5.2 Lower and Upper Brainstem

Lesions of the lower brainstem reduce and delay BAEP waves starting with wave II or III; they may increase IPLs I–III and I–V. Lesions of the upper brainstem reduce and delay waves IV and V; they may increase IPLs III–IV and I–V but spare IPL I–III. However, a clear distinction between these defects and a precise localization of the level of the lesion within the brainstem is often impossible.

   The side of unilateral brainstem lesions may be identified by comparing BAEPs to stimulation of each ear. A unilateral lesion may cause abnormalities only in the BAEP elicited by stimulation of the ipsilateral ear and recorded on both sides of the head.

   Brainstem lesions may also cause abnormalities of other AEPs, but these abnormalities do not generally help to localize the lesions to the stem.

### 12.5.3 Subcortical and Cortical Cerebral Connections

Although the BAEP waves VI and VII, like the peaks of the MLAEP and those of the LLAEP, are generated by subcortical and possibly cortical auditory connections, they are quite variable and have not been clearly related to specific structures. Therefore, they have not proved useful in detecting and localizing lesions in the cerebral parts of the auditory pathways.

### 12.5.4 Audiologic Strategies: Hearing Loss Due to Peripheral Defects

AEPs may be used to evaluate hearing ability. In principle, electric response audiometry (ERA)

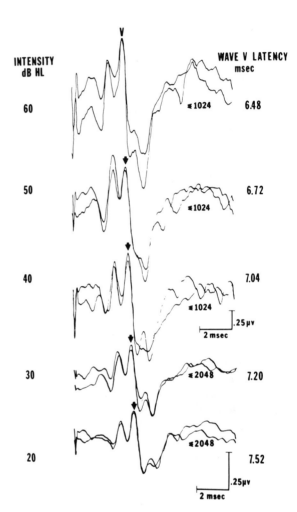

FIGURE 12.3.   *Normal BAEPs to click stimuli of decreasing intensity in a 13-month-old infant. Wave V (arrows) increases in latency (numbers to the right of tracings) and decreases in amplitude with decreasing stimulus intensity (note higher sensitivity for the last two tracings). Numbers below the ends of the tracings indicate numbers of responses collected for each BAEP. Two BAEPs are superimposed for each stimulus intensity. Recordings between vertex and mastoid on the side of stimulation. Positivity at the vertex is plotted upward. From Mokotoff et al. (Arch Otolaryngol 103:38, 1977) with permission of the authors and the American Medical Association.*

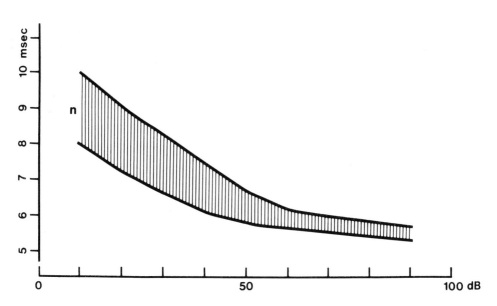

FIGURE 12.4. *Normal range of latency-intensity curves derived from BAEPs to click stimuli of different intensity in a group of audiologically normal adults. Horizontal axis: Intensity of the click stimulus in dB nHL. Vertical axis: Latency of wave V in msec.*

attempts to detect and to classify hearing losses by determining the threshold stimulus intensity that produces a visible AEP and by measuring the latency of AEPs produced by higher stimulus intensities. ERA may also be used to select hearing aids.

BAEPs are now used most widely for ERA. A common test procedure first applies a click stimulus of 60–70 dB nHL or of 110–120 dB peSPL to elicit a BAEP. Wave V is used as the indicator because it has the lowest threshold and highest amplitude of BAEP waves. If wave V can be distinguished at this level, stimuli of 20 or 30 dB nHL, or 75 dB peSPL, are used next (Figure 12.3). If wave V appears at this lower level, the search for the threshold is discontinued in most laboratories because this finding is consistent with normal hearing and further definition of the threshold is without practical audiological importance. If no BAEP can be distinguished either at the initial 60- or 70-dB nHL intensity or at the 20- or 30-dB level, averages to stimuli increasing in intensity by 10 dB are obtained until the threshold of wave V is found or a maximum of about 90 or 100 dB nHL is

reached. Latencies of wave V at different intensities may be used to plot latency-intensity curves showing the decrease of latency of wave V obtained from increasing the stimulus intensity (Figure 12.4). Separate normal latency-intensity curves must be used for different age groups.

ERA is limited by two problems: (1) AEP thresholds are difficult to determine and exceed the subjective hearing level to an extent that varies among subjects; (2) The most effective stimuli, namely, clicks and brief tones, test only the upper part of the audible frequency range. Lower frequencies, especially those of the speech range, can not be evaluated very effectively with the sudden and brief stimuli required for BAEPs and MLAEPs.

The correlation between the presence or absence of BAEPs and hearing ability is not perfect. Although most subjects showing BAEPs can hear, mild hearing defects may escape detection when only strong stimuli are used. Furthermore, low frequency hearing losses cannot be detected with stimuli of predominantly high frequency content such as clicks. On the other hand, BAEPs may be abnormal, especially in their late peaks, in subjects who can hear normally. This is common in patients with multiple sclerosis and is due presumably either to marked desynchronization of impulses or to abolition of impulses in some parts of the auditory system that may be critical for the production of BAEPs but not for hearing. Very rarely are BAEPs abolished without abolition of hearing. BAEPs may be partly mediated by auditory brainstem pathways that are not essential for hearing but subserve such other functions as the spatial localization of sound and the interaural discrimination of pitch and loudness.

# 13

# The Abnormal BAEP

## 13.1 CRITERIA DISTINGUISHING ABNORMAL BAEPS

### 13.1.1 Neurologic Applications

#### 13.1.1.1 Absence of all BAEP waves

The absence of waves I–V is established by a careful search for BAEP waves that includes the use of high stimulus intensities and of numbers of responses sufficient for a good reduction of residual noise in the recording. Technical problems must be excluded.

#### 13.1.1.2 Absence of all waves following wave I or wave II

Wave I must be carefully searched for in tracings showing mainly residual noise because preservation of wave I has important clinical implications.

#### 13.1.1.3 Abnormal prolongation of IPLs I–V, I–III, and III–V

The IPL I–V is calculated by subtracting the peak latency of wave I from the peak latency of

wave V or of the wave IV–V complex. If wave I can not be clearly distinguished in recordings between vertex and stimulated ear, the negative wave preceding wave II may be more clearly defined and should be used for IPL measurements. This wave may also be used for IPL measurements in recordings between vertex and nonstimulated ear that do not contain wave I. Normative data for these measurements should be collected.

IPLs I–III and III–V are determined by subtracting the latency of wave I from that of wave V and by subtracting the latency of wave III from that of wave V. An increase of one or both of these IPLs, usually associated with an increase of IPL V, may help to further localize the underlying lesion.

#### 13.1.1.4 Absence of wave V or abnormal decrease of the amplitude ratio of waves V and I

Peak-to-peak amplitude measurements of wave I and wave V or of the IV–V wave complex are used to calculate the amplitude ratio of wave V and wave I. This ratio is a more reliable measure than the amplitude of wave V itself and should be determined routinely except in cases of complete absence of wave V. The evaluation of the ratio is complicated by its dependence on click intensity and its variation with hearing loss. Because the ratio has a non-Gaussian distribution in the normal population, the normal range must be defined in terms other than those of mean and standard deviation.

### 13.1.1.5 Interaural latency difference of IPL I–V

The IPLs I–V to stimulation of each ear may differ to an extent that exceeds the normal range for this interaural latency difference, even if neither IPL itself exceeds the normal range for IPL I–V.

### 13.1.1.6 Abnormal increase of latency to rapidly repeating stimuli

The latency of wave V at stimulus rates of 60–100/sec is usually longer than that at 10–30/sec. An increase of this rate-dependent latency shift beyond the normal range suggests a lesion. This criterion is not used for clinical interpretation in most laboratories.

### 13.1.1.7 Abnormal peak latency

Because measurements of IPL are relatively immune to peripheral hearing loss, they are preferred to measurements of peak latencies in neurologic studies. However, the latency to the peak of wave I is useful in evaluating peripheral lesions. Furthermore, peak latency of waves V and III must be used instead of IPLs in instances where wave I and other waves preceding wave V can not be distinguished.

## 13.1.2 Audiological Applications

### 13.1.2.1 Abnormally high threshold

An increase of BAEP threshold above 30 dB nHL or 75 dB peSPL is abnormal.

### 13.1.2.2 Abnormally long latency of wave V with normal IPL I–V

In normal subjects of any age past infancy, the latency of wave V at 60 dB nHL does not exceed 6 msec. The peak latency of wave V is used in some laboratories to estimate hearing loss. This is justified only if the IPL I–V at the same stimulus intensity is not prolonged because such a prolongation may indicate a brainstem lesion.

### 13.1.2.3 Abnormal latency-intensity curve

The absolute value of latency of wave V and the slope of the latency-intensity curve depend on the subject's age and vary among laboratories. At any age past infancy, the decrease of latency with increasing stimulus intensity should not exceed 40 μsec/dB at middle intensities and about 30 μsec/dB at higher intensities of 60–70 dB HL. To determine the slope reliably, measurements should be made over a range of at least 30 dB.

## 13.2 GENERAL CLINICAL INTERPRETATION OF ABNORMAL BAEPS

## 13.2.1 Neurological Disorders

### 13.2.1.1 Absence of all BAEP waves

Inability to record waves I–V on stimulation of either ear may be due to various technical problems such as lack of an effective stimulus, faulty synchronization between stimulator and averager, or defects of recording electrodes, electrode connections, amplifiers, or averaging channels (Table 13.1). Absence of waves I–V on stimulation of one or both ears may be due to severe conductive or sensorineural hearing loss which should be further investigated audiologically. A neurological cause for the complete absence of waves I–V is a lesion of the distal acoustic nerve. Only in a few cases of absent BAEPs is hearing not abolished.

### 13.2.1.2 Absence of all waves following wave I or wave II

Preservation of initial waves indicates that the absence of later waves is not due to technical problems or peripheral lesions and therefore provides clear evidence for a retrocochlear lesion involving the proximal acoustic nerve or the pontomedullary region of the brainstem. Bilateral absence of all waves after wave I may be seen in brain death.

TABLE 13.1.   Clinical interpretation of abnormal BAEPs

| *Abnormality* | *Interpretation* |
| --- | --- |
| Absent bilaterally | Bilateral acoustic nerve lesions; brain death; rule out technical problem |
| Low amplitude or increased latency of entire BAEP bilaterally | Peripheral hearing loss; acoustic nerve lesion; rule out reduced stimulus intensity |
| Absent; other side normal | Unilateral cochlear or acoustic nerve lesion |
| Absent peaks after a normal wave I, other side normal | Unilateral proximal acoustic nerve or pontomedullary lesion |
| Increased latency of wave I and subsequent waves; IPL III–V normal | Peripheral hearing loss or acoustic nerve lesion |
| Absent wave V or decreased amplitude ratio V/I | Ipsilateral lower or upper brainstem lesion |
| Increased IPLs I–III and III–V | Ipsilateral lower and upper brainstem lesion |
| Increased IPL I–III; normal III–V | Ipsilateral lower brainstem lesion, between acoustic nerve and lower pons |
| Increased IPL III–V; normal I–III | Ipsilateral upper brainstem lesion, between lower pons and midbrain |
| Abnormal increase of wave V latency with rapidly repeating stimuli | Suspect ipsilateral brainstem lesion |
| Increased BAEP threshold | Suspect peripheral hearing loss or distal acoustic nerve lesion |
| Shift of latency-intensity curve upward but parallel to normal curve | Conductive hearing loss |
| Shift of latency-intensity curve upwards predominantly at low intensities | Sensorineural hearing loss |

### 13.2.1.3 Absent wave V or decreased amplitude ratio of waves V and I

Selective reduction or abolition of wave V often indicates a midbrain lesion.

### 13.2.1.4 Increased latency of all BAEP waves

A delay of all recorded BAEP waves may be due to a reduction in stimulus intensity caused by technical problems. Hearing loss, especially conductive, may have the same effect. However, a lesion of the distal acoustic nerve may also delay the entire BAEP. The IPL I–V should be normal in those instances. An additional central lesion must be suspected if this IPL is also increased. In the absence of a measurable wave I, a delay of waves III and V may be attributable to a peripheral problem if the IPL III–V is normal.

### 13.2.1.5 Increased IPL I–V or increased interaural difference of IPLs I–V

Abnormal separation between wave I and V usually indicates a central lesion. The lesion may be further localized by measurements of IPLs I–III and III–V. The increased IPL I–V is of such diagnostic importance that it is often searched for by comparing the IPL I–V on both sides; an abnormal difference may indicate a lesion on the side of the longer IPL I–V. Even in cases where wave I and other waves preceding wave V can not be distinguished, the finding of a normal peak latency of wave V is very useful because it practically excludes both central and peripheral lesions.

The rule that an increased IPL I–V means a central lesion has several exceptions. Acoustic nerve lesions may cause abnormalities beginning with wave I and thus resemble peripheral lesions. Both a reduction of stimulus intensity and a change of stimulus polarity can increase the IPL I–V. Patients with recurrent middle ear disease may develop increased IPLs. Peripheral hearing losses, causing a reduction of stimulus intensity, may lengthen IPL I–V if patients are tested at stimulus intensities below that used as the laboratory standard. Although mild conductive hearing defects can be compensated for by increasing the stimulus intensity to a standard level above the patient's hearing threshold, it may be impossible to overcome the effect of severe hearing losses. If a peripheral hearing loss reduces the effective stimulus intensity relative to the patient's hearing ability, the IPL may be found to be increased in comparison with normative values for standard stimulus intensities. The situation is even more complicated in high-frequency hearing losses which increase the latency of wave I without increasing the latency of wave V to the same degree, thereby reducing the IPL I–V and theoretically decreasing the ability to detect central lesions which increase the IPL I–V. Another problem arises if wave I, having higher threshold than wave V, is absent from a tracing that shows an abnormally prolonged latency of wave V; this makes it impossible to use IPL I–V as a criterion for the distinction between central and peripheral lesions. The interpretation of BAEPs for neurological applications must take into account these audiological problems to avoid misinterpretations. Audiological evaluation should be obtained in case of any doubt about the central cause of BAEP abnormalities. Such doubt is especially likely to arise in cases showing absence of all BAEP waves or of all waves after wave I.

### 13.2.1.6 Abnormal increase of latency of wave V to rapidly repeating stimuli

An abnormal shift of wave V latency with increasing stimulus rate may suggest a brain-stem lesion. Rate-dependent latency shifts of earlier BAEP peaks can usually not be measured reliably because these peaks rapidly lose amplitude at high stimulus rates.

### 13.2.2 Audiological Disorders

In hearing loss due to cochlear and middle ear disorders (Table 13.1), the BAEP may show an increased threshold and an increased latency of wave I and all subsequent waves. Plotting of the latency-intensity curve can distinguish conductive, sensorineural, and other hearing losses.

## 13.3 NEUROLOGICAL DISORDERS THAT CAUSE ABNORMAL TRANSIENT BAEPS

BAEPs are most useful in the diagnosis of:

- multiple sclerosis,
- extramedullary and intramedullary brainstem tumors,
- coma due to metabolic and structural lesions, and
- several other disorders in which BAEPs have been reported to be abnormal.

In infants, BAEPs are usually evaluated for both neurological and audiological purposes to search for lesions in the central auditory pathway and for peripheral hearing defects. BAEPs with audiometry are increasingly used for monitoring patients receiving ototoxic chemotherapy.

### 13.3.1 Multiple Sclerosis

Abnormal BAEPs are found in many patients with MS; abnormalities occur most often in patients who have a diagnosis of definite MS and in patients who show clinical signs of brainstem involvement. However, abnormal BAEPs are found often enough in patients with probable or possible MS who do not show evidence of brainstem lesions to make the test diagnostically valuable.[6,20] The BAEP may suggest the diagnosis of MS in patients who present with retrobulbar neuritis or transverse myelitis.

No particular BAEP abnormality is specific for MS: The type of abnormality varies with

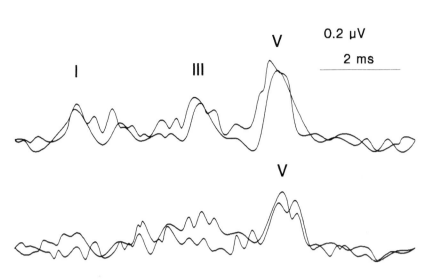

FIGURE 13.1. *Abnormal BAEPs with increased IPL I–V ipsilateral to the stimulated ear and increased peak latency on the other side. The wave between I and V in the ipsilateral BAEP probably represents wave III and suggests an increased IPL I–III. This would be consistent with a lesion in the lower brainstem. This 49-year-old man has MS manifest by progressive spastic paraparesis of two years and recent onset of bilateral cerebellar signs.*

the location and the severity of the demyelinating lesions. Figure 13.1 shows an increased IPL I–V and, probably, an increase of IPL I–III, suggesting a lesion in the lower brainstem. Figure 13.2, from another patient, shows nearly complete abolition of wave V with preserved

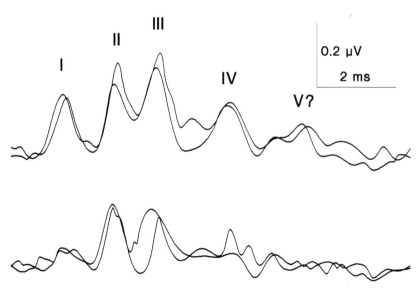

FIGURE 13.2. *Abnormal BAEP to stimulation of the left ear showing only a questionable wave V (second cursor) which, if used for measuring, indicates abnormal IPLs I–V and III–V, and abnormally low amplitude ratio V/I ipsilateral to the stimulated ear; in the contralateral recording, wave V is completely abolished. These findings suggest a lesion in the upper brainstem. This 27-year-old woman was suspected of having multiple sclerosis because of a right central scotoma, present for five years, and the recent onset of horizontal and rotatory nystagmus.*

waves I–III, suggesting a lesion in the upper brainstem. Such findings can help to distinguish central lesions from vestibular lesions producing vertigo and nystagmus and thus mimicking brainstem involvement. Sequential recordings of BAEPs in patients with MS show that a BAEP abnormality may increase or decrease with time, and may occasionally disappear completely. Although such fluctuations are often related to clinical changes, they have also been found to occur in clinically stable patients and are therefore not useful for serial monitoring of changes in clinical condition.

The effectiveness of the BAEP in detecting abnormalities in patients with MS has been compared with that of other EPs. BAEPs have generally been reported to be less effective than VEPs to pattern reversal and SEPs.

### 13.3.2 Extramedullary and Intramedullary Brainstem Tumors

#### 13.3.2.1 Cerebellopontine angle tumors

Acoustic neurinomas and meningiomas impinging on the brainstem nearly always produce BAEP abnormalities at the time of diagnosis. False-negative results are very rare; false-positive results are more common because BAEP abnormalities are not specific and may result from damage to the proximal or distal parts of the acoustic nerve and from compression and

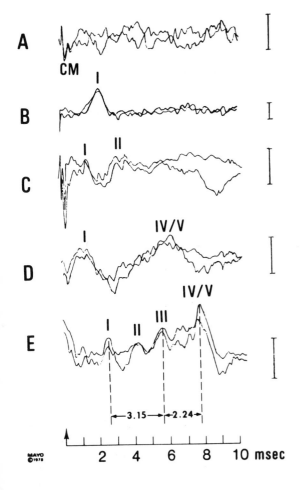

FIGURE 13.3.   *Five abnormal BAEP patterns in cerebellopontine angle tumors. (A) Absence of all BAEP waves and preservation of the cochlear-microphonic (CM) in a large intracanicular acoustic neurinoma extending into the cerebellopontine angle and compressing the brainstem. (B) Absence of waves after a delayed wave I in a small petrous ridge meningioma compressing the extracanicular portion of the eighth nerve and the pontomedullary junction of the brainstem. (C) Absence of waves after wave II with increased IPL I–II in a lesion near the entrance of the acoustic nerve into the brainstem. (D) Increased IPL I–V, decreased amplitude ratio (V/I) and absence of wave III in a large extracanicular neurinoma causing severe pontine compression and bilateral cranial nerve defects. (E) Increased latency of wave I and increased IPL I–III in a small cholestoma involving the eighth nerve at its entrance into the cerebellopontine angle. Click stimulation at 60 dB SL of the ear on the side of the tumor. Recording between vertex and stimulated ear. Positivity at the vertex is plotted upward. Calibration 0.1 µV. From Stockard et al. (In Electrodiagnosis in Clinical Neurology, ed. M. J. Aminoff, pp. 370–413, Churchill Livingstone, 1980) with permission of the authors and Mayo Clinic.*

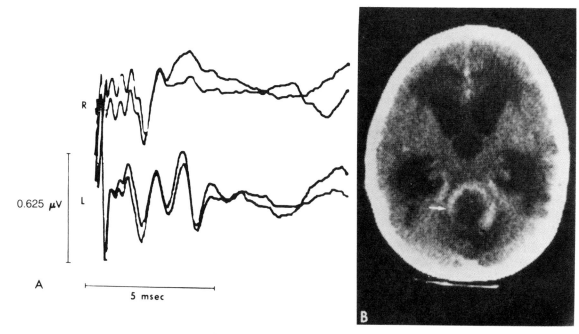

FIGURE 13.4.   *BAEPs in a patient with intramedullary brainstem ependymoma. The BAEP to stimulation of the right ear (R) shows only waves I and II. The BAEP to stimulation of the left ear (L) shows waves I, II, and III with normal IPLs. Click stimuli at 80 dB nHL. Recordings between vertex and stimulated ear. Negativity at the vertex is plotted upward; the BAEP waves point down. CT scan shows an enhancing lesion filling the fourth ventricle and situated to the right of the midline. From Nodar et al. (Laryngoscope 90:250, 1980) with permission of the authors and Laryngoscope Company.*

distortion of the brainstem. The high incidence of abnormal findings has made the BAEP, in combination with the MRI scan, the most powerful tool in the diagnosis of acoustic neurinomas.

A frequent BAEP abnormality in cases of acoustic neurinomas is the absence, increased latency, or increased duration of wave I (Figure 13.3 *A, B*). Subsequent waves are usually severely abnormal or absent, although they may be preserved, but delayed (Figure 13.3 *C, D*). Wave V often has an increased peak latency in comparison either with normal ears or with the opposite ear of the same subject. If waves I and III are preserved, the IPL I–III is often prolonged as a result of pontine compression; this prolongation is a very sensitive indicator of cerebellopontine angle tumors (Figure 13.3 *E*). An increased IPL III–V may be found on recording from the opposite side, probably as a result of distortion of the lower midbrain. The BAEP may normalize after surgical removal of the tumor.

### 13.3.2.2 Intramedullary brainstem tumors

Wave I, generated by the acoustic nerve, is usually preserved, whereas later waves are very often abnormal (Figure 13.4). Waves III, IV, and V may be absent; if present, they may show increased latency and decreased amplitude. The IPL I–III is increased if the tumor is located in the pontomedullary region or pons, and the IPL III–V is increased if it involves the pontomesencephalic junction or midbrain. Several IPLs of the BAEPs to stimulation of each ear may be increased by diffuse brainstem tumors; the BAEP is a very sensitive indicator of infiltrating pontine gliomas.

### 13.3.2.3 Surgical monitoring

BAEPs have been useful for monitoring the condition of the acoustic nerve and brainstem during surgery, especially since the BAEP is not affected by ordinary anesthetics. BAEPs have therefore been used in various types of operations of the posterior fossa. These include surgery for acoustic neuromas, microvascular decompression, brainstem tumors, and basilar or vertebral artery aneurysms.

Loss of BAEP waveform or increased latencies suggests dysfunction of the auditory pathways which may be relieved by release of retraction or alteration of operative manipulation. During surgery for acoustic neuroma, the most common changes are loss of waves II–V or increased I–III interpeak latency. There is a good but not perfect correlation between deterioration of intraoperative BAEP and subsequent hearing deficit. The role of intraoperative BAEP in predicting deficit from basilar aneurysm surgery is more controversial.

### 13.3.3 Coma

BAEPs may help to distinguish coma due to structural lesions from coma due to metabolic, toxic, and other potentially reversible encephalopathies: Structural lesions may cause abnormal BAEPs, whereas metabolic and toxic encephalopathies are generally associated with normal BAEPs unless they cause irreversible damage. However, normal BAEPs can not definitely exclude the possibility of structural damage in comatose patients, especially damage involving higher cerebral structures.

#### 13.3.3.1 Coma due to structural lesions: Assessment of survival

In comatose patients with head injuries, the BAEP has moderate prognostic value; it is less useful than other EPs. Survivors have been reported to show normal BAEPs after several months. The BAEP is abnormal in the majority of patients with coma due to meningoencephalitis, but the BAEP is not accurate for prognosis. In spontaneous intracerebral hemorrhage, the BAEP has been found to indicate brainstem damage and to predict outcome.

Widespread postanoxic or posttraumatic cerebral damage is often associated with abnormal BAEPs (Figure 13.5), but may be associated with normal BAEPs, especially in patients who show clinical or laboratory evidence for widespread damage to cortical, subcortical, and diencephalic structures but have preserved brainstem function. In contrast, patients in coma due to pontomesencephalic encephalomalacia show abnormalities of early waves corresponding with the level and side of the lesion or show no recognizable waves at all.

Patients with sufficient brainstem damage to fulfill the criteria of cerebral death show no BAEP waves after wave I or, occasionally, after wave II. The BAEP can suggest the diagnosis of cerebral death only if wave I is preserved because the complete absence of BAEPs may be due to deafness caused by peripheral auditory problems, lesions of the acoustic nerve, or malfunctioning of stimulating and recording equipment. The disappearance of the BAEP does not necessarily indicate brain death but has occasionally been observed to be transient even in cases of structural damage. On the other hand, preservation of the BAEP may help to refute the diagnosis of cerebral death.

Patients with pontine hemorrhage with absence of all waves after II all died.[34] In the same study, two patients with normal amplitudes of waves I–V on at least one side both survived.

#### 13.3.3.2 Coma due to metabolic and toxic encephalopathies

In coma due to metabolic derangements such as uremia, hepatic failure, and diabetic ketoacidosis, or coma due to CNS-depressant drugs, the BAEP is nearly always normal. Anesthesia with halothane and sodium thiopental does not affect the BAEP. Exceptions are enflurane anesthesia and imipramine overdose, both of which increase IPLs. The BAEP may persist even if the EEG shows electrocerebral silence and the clinical condition indicates abnormal brainstem function. However, the BAEP may be abnormal or absent in patients who have preexisting hearing loss and in patients who suffer structural brainstem damage in the course of metabolic or toxic coma, although increased IPLs have been reported in irreversible hepatic encephalopathy without anatomically demonstrable brainstem damage.

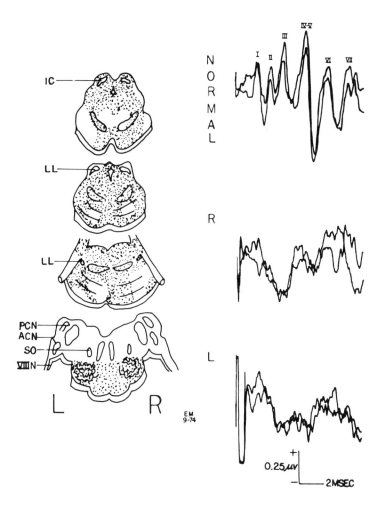

FIGURE 13.5. *BAEPs in a 3-month-old patient with postanoxic encephalopathy. Top tracings show normal adult BAEPs. Middle and bottom tracings show BAEPs to stimulation of the right and left ear of the patient; waves I have normal latency for the patient's age, but later waves, although visible, can not be identified by comparison with normal waves. Stimulation with clicks of 75 dB nHL. Recordings between vertex and stimulated ear. Positivity at the vertex is plotted upward. Two tracings are superimposed for each condition. Damaged areas are indicated by stippling in the diagram at the left. From Starr and Hamilton (EEG Clin Neurophysiol 41:595, 1976) with permission of the authors and Elsevier Scientific Publishers Ireland Ltd.*

A fall of body temperature to 32°C or less may occur in metabolic or toxic coma and is another possible cause of abnormal BAEPS in these disorders. In infants, metabolic disorders such as phenylketonuria, Leigh's disease, and maple syrup urine disease may lead to coma associated with BAEP abnormalities that are probably due to faulty myelination.

## 13.3.4 Other Disorders

The BAEP has been reported to be abnormal in many disorders involving the brainstem. Although the BAEP may be of diagnostic use in only a few of them, these associations must be considered in the differential diagnosis to avoid misinterpretation of abnormal BAEPs.

### 13.3.4.1 Degenerative and other diseases of unknown cause involving the brainstem

Abnormal BAEPs have been reported in several disorders:

- olivopontocerebellar degeneration (Figure 13.6)
- Friedreich's ataxia
- hereditary cerebellar ataxia
- Alzheimer's disease
- Parkinson's disease

- hemifacial spasm
- myotonic dystrophy
- hereditary sensory neuropathy
- Dejerine-Sottas disease
- albinism
- Leber's optic atrophy
- hydrocephalus

Normal BAEPs were reported in:

- supratentorial lesions
- spinocerebellar degenerations other than those listed above
- progressive supranuclear palsy
- Jakob-Creutzfeldt disease
- Huntington's chorea
- amyotrophic lateral sclerosis
- Ménière's disease
- labyrinthitis and vestibular neuronitis
- subacute sclerosing panencephalitis
- most cases of palatal myoclonus
- Tourette's syndrome
- during seizures induced by electroconvulsive therapy
- precocious puberty
- narcolepsy
- schizophrenia and depression
- alternating hemiplegia of childhood
- migraine

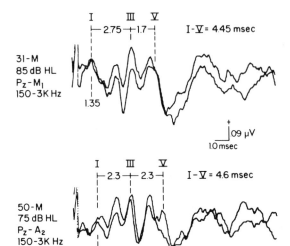

FIGURE 13.6.   *Abnormal BAEPs from two patients with olivopontocerebellar degeneration showing abnormally prolonged IPL I–III (top) and IPL III–V (bottom). Two tracings are superimposed for each BAEP. Recordings between vertex and stimulated ear. Positivity at the vertex is plotted upward. From Lynn et al. (Semin Hear 4:375, 1983) with permission of the authors and Thieme-Stratton Inc.*

Several conditions are associated with significantly increased interpeak latencies when a certain population of patients is considered, for example, pregnant women and people who suffer migraines.[28,84] However, on an individual basis, the prolongation is not sufficient to interpret these studies as abnormal. The reader should be cautious regarding interpretation of small differences in latencies among populations of patients.

### 13.3.4.2 Toxic and metabolic disorders without coma

Drinking alcohol produces an acute increase of the latency of BAEP waves following wave I, which is probably not explained by hypothermia. IPLs may also be increased in chronic alcoholics even after abstinence. A reversible increase of IPL I–V was observed in two alcoholics clinically diagnosed as having central pontine myelinolysis. The BAEP remains normal in Wernicke's encephalopathy.

Most sedatives and anesthetics do not affect BAEPs even in doses causing loss of consciousness. A slight increase of IPLs was found in patients with phenytoin levels greater than 20 μg/ml; this probably does not significantly interfere with the ability of the BAEP to detect brainstem lesions in patients on therapeutic phenytoin. Toluene sniffing was found to abolish BAEPs following wave II in two patients with cerebellar signs and dementia. Treatment with aminoglycosides may produce BAEP changes.

IPLs may be slightly prolonged in diabetics even at times when the blood sugar is not increased. IPL I–III is particularly affected.[59] BAEPs may be abnormal in Leigh's disease, Wilson's disease, and in hyperglycemia. Vitamin $B_{12}$ deficiency caused abnormal BAEPs in one but not another study. Only a few patients with chronic renal failure had delays of wave V or increased IPLs. Hypoxia due to breathing air with 9% to 13% $O_2$, or hypercapnia due to breathing air with 7.5% to 10% $CO_2$, did not affect the BAEP.

Patients with Wilson's disease may have abnormal BAEPs, especially if they have neurologic deficits. Severely affected individuals had increased IPL III–V, indicating CNS dysfunction.[15]

### 13.3.4.3 Strokes

Brainstem strokes may cause abnormal BAEPs. The locked-in syndrome may be associated with an increased IPL III–V or other BAEP abnormalities, but normal BAEPs have also been reported in this syndrome and in Wallenburg's lateral medullary syndrome. Cortical deafness due to bilateral vascular lesions of the temporal lobe was associated with normal BAEPs.

Brainstem transient ischemic attacks may be associated with abnormal BAEPs during the deficit, but normalization when the patient is asymptomatic.[29]

### 13.3.4.4 Vascular malformations of the posterior fossa

Arterial and venous malformations associated with hemifacial spasm, trigeminal neuralgia, and facial paresis may cause peak and interpeak latency abnormalities similar to those occurring in cerebellopontine angle tumors, although normal BAEPs have also been reported in trigeminal neuralgia.

### 13.3.4.5 Leukodystrophies

Abnormal BAEPs were found in patients with Pelizaeus-Merzbacher disease, adrenoleukodystrophy, and metachromatic leukodystrophy, but not in patients with subacute sclerosing panencephalitis and with gray matter degenerations such as Batten's disease and Hallervorden-Spatz disease.

### 13.3.4.6 Postconcussion syndrome

Prolonged IPLs after head injuries suggest residual organic brainstem damage.

### 13.3.4.7 Mental retardation

Adult patients with severe mental retardation due to Down's syndrome or to unknown causes often show BAEP abnormalities indicating central damage or profound hearing loss. The latency of wave V is shorter and the latency-intensity curve is steeper in infants with Down's syndrome than in normal subjects.

### 13.3.4.8 Sleep apnea and sudden infant death syndrome

Sleep apnea in adults is associated with normal BAEPs unless it is a symptom of a massive brainstem lesion. A few BAEP investigations were unable to reliably discriminate infants at risk for the sudden infant death syndrome, although another study found an increased IPL I–V in a group of premature babies with apnea.

### 13.3.4.9 Neurological disorders in infants

Even though BAEPs are used in infants mainly to evaluate hearing, brainstem problems may be detected by prolonged IPLs and other abnormalities. Infants with perinatal problems, such as complications during pregnancy, birth injury, respiratory distress, intraventricular hemorrhage, and apneic syndrome, show an abnormal BAEP development. Asphyxic damage leading to neurologic handicaps is characterized by normal BAEPs. Some children with psychomotor retardation or "minimal brain dysfunction" have BAEP findings suggesting central abnormalities. Although some patients with infantile autism show BAEP abnormalities suggesting profound hearing loss, others show evidence for central problems. BAEPs were severely abnormal in some infants with Gaucher's disease. In children recovering from bacterial meningitis, BAEPs can detect sensorineural hearing loss.

### 13.3.4.10 Infections

Meningitis may produce increased I–III interpeak latencies due to damage to the eighth nerve as it passes through the subarachnoid space. Both bacterial and aseptic meningitides may produce this abnormality. After recovery from meningitis, abnormalities in BAEP correlate with developmental deficits; the most sensitive indicator of neurological dysfunction was reduced wave V amplitude, in one study.[43]

A small proportion of patients with AIDS have abnormalities on BAEP. Other EP modalities are more clinically useful.

## 13.4 AUDIOLOGICAL DISORDERS THAT CAUSE ABNORMAL BAEPS

### 13.4.1 Effects of Hearing Loss on Threshold and Latency

#### 13.4.1.1 Effect of hearing loss on threshold

Hearing loss increases the threshold of the BAEP, specifically that of wave V if the loss involves the frequencies of 1–4 kHz through which click stimuli exert their effect. Although the threshold of wave V may be only 5–10 dB above the hearing threshold in many adults and within 10–20 dB above the adult normal hearing level in many normal infants, increases of threshold of less than 30 dB above the normal hearing level can not be taken as evidence of abnormal hearing. Even then, the method is not absolutely reliable: A few subjects with behavioral evidence for hearing and without indication for brainstem lesions have no recordable BAEP, suggesting that the pathway tested by BAEPs is not entirely the same as that needed for hearing. On the other hand, a normal threshold to click stimulation does not guarantee normal hearing, especially since this stimulus tests hearing at frequencies of mainly 2–4 kHz. For the same reason, it is not possible to determine the degree of hearing loss from the degree of threshold elevation. But even with these limitations, BAEPs to clicks are probably the best currently available electrophysiological test of hearing ability. They can be used to detect hearing losses in infants, autistic and difficult-to-test children, and retarded children and adults. In contrast, nonorganic hearing deficits are characterized by much lower BAEP thresholds than predicted by behavioral hearing tests.

#### 13.4.1.2 Effect of hearing loss on latency

Hearing loss increases the latency of wave V above the normal level. This increase has occasionally been used to estimate hearing loss. However, the relationship between these two variables is not simple. Central lesions must be excluded by demonstrating a normal IPL I–V at the same stimulus intensity used. Even then, the degree of latency change depends on the type of hearing loss and is not necessarily proportional to the stimulus intensity.

#### 13.4.1.3 Effect of hearing loss on the latency-intensity curve

Hearing loss can be further characterized by plotting the latency of wave V against the stimulus intensity and comparing the latency-intensity curve with a plot of the normal range. This method may help to distinguish conductive from sensorineural hearing loss.

### 13.4.2 Conductive Hearing Loss

#### 13.4.2.1 Effect on amplitude and latency

Conductive hearing loss interferes with the conduction of sound waves from the ear canal to the cochlea and therefore acts like a reduction of stimulus intensity. The reduction of the click stimulus intensity produces a BAEP of lower amplitude and longer latency of all waves. At low stimulus intensities, the latency of wave I is more increased than that of other waves in conductive hearing loss, so that IPLs I–V and I–III are shortened. Compensation for these effects, if caused by moderate conductive hearing losses, can be made by increasing the stimulus intensity. However, in many cases it may be impossible to completely overcome the hearing defect by increasing the stimulus intensity. In these cases, some laboratories attempt to derive neurologically useful information by correcting latencies for the degree of hearing loss or by using midline loudness matching to determine the stimulus intensity to be used for the ear with decreased hearing. The general rules relating conductive hearing loss to BAEPs do not apply to conductive hearing loss caused by ossicular chain disorders.

Because bone conduction is not affected in conductive hearing losses, some investigators have demonstrated the discrepancy between air and bone conduction by comparing BAEPs to earphone click stimulation with BAEPs produced by a bone vibrator stimulator or by recording BAEPs to clicks masked by bone-conducted noise. Impedance audiometry is probably at least as effective in diagnosing conductive defects as any BAEP test.

### 13.4.2.2 Effect on latency-intensity curve

Conductive hearing defects increase the latency of wave V equally over the entire range of intensities and therefore shift the latency-intensity curve by an amount equal to the hearing loss; a parallel shift of this sort is therefore characteristic of conductive defects (Figure 13.7). The possibility of a central defect must be excluded by ascertaining that the IPL I–V is normal.

### 13.4.3 Sensorineural Hearing Loss

#### 13.4.3.1 Effect on amplitude and latency

In sensorineural hearing loss, the BAEP amplitude is decreased, at least at moderate stimulus intensities, and the amplitude ratio V/I is usually increased. The latency of wave V is increased, in keeping with the degree of hearing loss at 4 kHz. However, wave I, if visible, is at least equally increased in latency, causing a normal or even abnormally short IPL I–V.

BAEP recordings that include the cochlear microphonic may be used to subdivide sensorineural hearing losses into neural hearing loss in which wave I is absent while the cochlear microphonic is preserved and sensory hearing loss in which both wave I and cochlear microphonic are absent.

#### 13.4.3.2 Effect on latency-intensity curve

Many causes of high-frequency sensorineural hearing loss show the greatest deviation from normal latency and amplitude at low

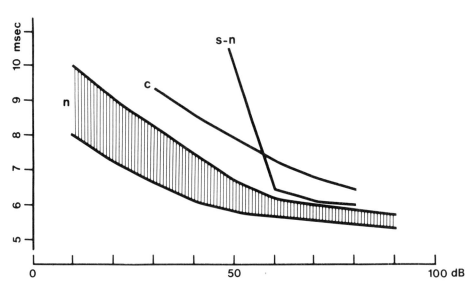

FIGURE 13.7. *Latency-intensity curves in cases of conductive hearing loss (c) and sensorineural hearing loss (s-n) as contrasted with the normal range. Horizontal axis: Intensity of the click stimulus in dB nHL. Vertical axis: Latency of wave V in milliseconds.*

stimulus intensities: The more the stimulus strength exceeds the threshold, the less the disparity between the abnormal and the normal curve; at high intensities, the latency may be normal. This causes a characteristic steepening of the latency-intensity curve which becomes L-shaped, having a very steep slope of up to 240 µsec/dB near threshold and a less steep slope of under 30 µsec/dB at higher intensities (Figure 13.7). The increased threshold and rapid decrease of abnormally prolonged latency

of wave V is often found in patients who have a hearing defect characterized by recruitment, that is, a hearing deficit that is maximal near threshold and decreases with increasing loudness; this kind of deficit is common in many cases of cochlear lesions, but rare in retrocochlear lesions. Electric and behavioral recruitment are characteristic of many disorders causing high-frequency sensorineural hearing loss, for instance, Ménière's disease and presbyacousis.

# 14

# Other AEPs

## 14.1 THE SLOW BRAINSTEM AEP

Stimulation with tone pips of gradually increasing amplitude produces a single wave that has a longer duration and peak latency than the BAEP produced by clicks;[49,50] peaks resembling those of the BAEP may be superimposed on the slow wave at high stimulus intensities (Figure 14.1). This slow wave has positive polarity at the vertex and is most likely equivalent to a slow negative wave at about 10 msec (SN10) recorded by other authors using higher settings of the low frequency filter. Like the BAEP, this slow potential is probably generated by the brainstem.

The slow brainstem potential is of interest to audiologists because its threshold is close to the subjective hearing threshold and because it seems to be fairly specific for even relatively low tonal stimulus frequencies of 500 Hz and above.

## 14.2 THE FREQUENCY FOLLOWING POTENTIAL

Stimulation with tone bursts of a frequency of 100–2,000 Hz and a duration of at least a few cycles, repeated at 10–15/sec, and recording between the vertex and ipsilateral or contralateral ear electrodes produce the frequency following potential (FFP) which consists of a sinusoidal rhythm having the frequency of the stimulus and riding on a sustained elevation of the baseline (Figure 14.2).[9] A transient BAEP, especially a wave V, may be distinguished during the first few cycles of the FFP. The FFP has irregular shape in some subjects; others show a response only at the onset of the stimulus. The FFP varies depending on electrode placement and may be partly obscured at ear electrodes by the cochlear microphonic and by postauricular responses. At a frequency of 500 Hz and an intensity of about 60–70 dB SL, the latency from the onset of the tone burst to the FFP is about 6 msec; it increases with decreasing tone frequency and intensity. The response to clicks given at the same frequency and a similar intensity as the waves of the stimulus tone differs from the FFP and shows superimposition of different AEP peaks.

The FFP is generally believed to be generated by brainstem structures that also produce the BAEP, perhaps mainly by structures near the inferior colliculus, although a peripheral origin or contribution has also been proposed.[81]

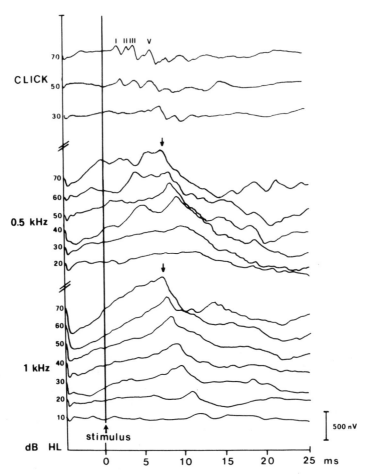

FIGURE 14.1.  *The normal slow brainstem AEP to tone pips of 0.5 and 1 kHz as contrasted with the BAEP to clicks. Stimulation with tone pips of decreasing intensity produces slowly rising AEPs which decrease in amplitude and increase in latency with decreasing stimulus intensity and can be distinguished to a stimulus level of about 20–30 dB HL. Stimulation with clicks of decreasing intensity induces BAEPs with decreasing numbers of peaks and decreasing amplitude; wave V can barely be distinguished at 30 dB HL. Broadband clicks are produced by electric pulses of 0.1 msec. Tone pips of 0.5 kHz have a duration of 6 msec and rise and fall times of 3 msec. Tone pips of 1 kHz have a 3-msec duration and 1.5-msec rise and fall times. Recording between forehead and mastoid on the stimulated side. Filter bandpass 0.2–2 kHz for BAEPs and 0.02–5 kHz for slow brainstem AEPs. Positivity at the forehead is plotted upward. From Maurizi et al. (Audiology 23:75, 1984) with permission of the authors and S. Karger AG.*

The audiometric use of the FFP is limited for two reasons: (1) The threshold of the FFP is much higher than the hearing threshold; (2) even though the FFP has the same frequency as the tone stimulus, it is not entirely specific for the stimulus frequency since FFPs to tone pips of relatively low frequency may be reduced by the simultaneous application of a constant masking sound of higher frequency and by high-frequency hearing defects. Low-frequency hearing losses are especially likely to escape detection. This has prevented the FFP from becoming used widely in electric response audiometry in spite of some encouraging initial results.

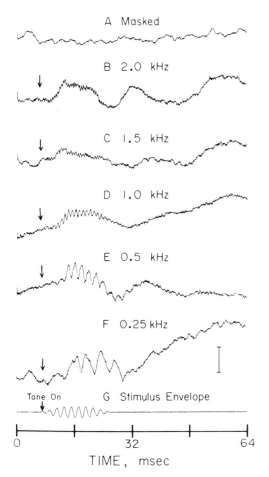

A Masked

B 2.0 kHz

C 1.5 kHz

D 1.0 kHz

E 0.5 kHz

F 0.25 kHz

Tone On    G Stimulus Envelope

0          32          64
TIME, msec

FIGURE 14.2. *The normal FFP to stimulation with tones of various frequencies. (A) Absence of a FFP to stimulation with tones of 0.5 kHz completely masked by simultaneous application of continuous white noise. (B–F) FFPs to stimulation with tones of 2, 11.5, 1, 0.5, and 0.25 kHz. (G) Tone stimulus of 0.5 kHz. Recording between vertex and stimulated ear. Positivity at the vertex is plotted upward. Vertical calibration bar indicates 0.5 μV for A–F. From Moushegian et al. (EEG Clin Neurophysiol 35:665, 1973) with permission of the authors and Elsevier Scientific Publishers Ireland Ltd.*

## 14.3 THE MIDDLE-LATENCY AEP

### 14.3.1 Methods of Stimulating and Recording

The subject's neck muscles must be completely relaxed because the slightest muscle tension

facilitates sonomotor responses that tend to obscure the MLAEP. Changes of attention, light sleep, and mild sedation do not affect the shape or threshold of the MLAEP.

Although clicks and high-frequency tone bursts are the most effective stimuli, the major audiological interest in the MLAEP lies in its potential to give frequency-specific information. Therefore, tone pips and filtered clicks are often used. However, the response seems to depend at least as much on the sudden onset of the stimulus as on its frequency. Stimuli with a rise time of 5 msec are more effective than stimuli with more slowly increasing amplitude, and those with rise times of over 25 msec produce unstable responses. Low-frequency tone bursts are less effective in eliciting MLAEPs than are filtered clicks. Stimuli may be repeated at rates of up to 15/sec without a change of response. At higher rates, amplitude begins to decrease.

Recordings are usually made between electrodes on vertex and ear or mastoid. Only one channel is used because MLAEPs to monaural stimuli recorded from the two sides of the head do not differ. Amplification requires a gain almost as high as that of the BAEPs because MLAEPs are only slightly larger than BAEPs. To avoid distortion of slow components of the MLAEP, the low frequency filter is usually set at about 10 Hz or less although a low filter of 25 Hz may speed up collection of MLAEPs. Higher frequency filter settings of less than 150 Hz may distort amplitude and latency of the MLAEP. The sweep length is usually about 100 msec. This results in a dwell time of about 100 μsec per point, or a sampling rate of at least 10 kHz, for the averager having 1,024 points per channel. Usually 1,000–2,000 responses are collected, although 512 responses or even fewer may be sufficient. With proper methods, the MLAEP may be recorded simultaneously with the BAEP.

### 14.3.2 The Normal MLAEP

Before the discovery of BAEPs, the MLAEP had been called early or early cortical AEP. It consists of several peaks of up to more than 1 μV which occur 10–50 msec after the stimulus (Figure 14.3).[63] They are often obscured by

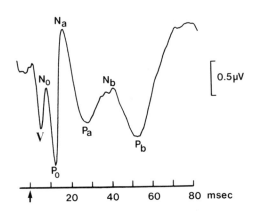

FIGURE 14.3.  *The normal MLAEP. This tracing is the grand average of MLAEPs from 35 normal subjects. Wave V of the BAEP preceded the MLAEP waves $N_0$, $P_0$, $N_a$, $P_a$, $N_b$, and $P_b$. Binaural click stimulation at about 75 dB SL (arrow). Recording between vertex and mastoid. Negativity at the vertex is plotted upward; BAEP wave V points down. Slightly modified from Robinson and Rudge (Brain 100:19, 1977) with permission of the authors and Oxford University Press.*

electrical activity generated in scalp and neck muscles. MLAEPs vary much more among laboratories and subjects than do BAEPs. Nevertheless, there is some agreement that the MLAEP may show up to five peaks of negative (N) and positive (P) polarity at the vertex: $N_0$ at 8–10 msec, $P_0$ at 10–13 msec, $N_a$ at 16–30 msec, $P_a$, usually the largest at 30–45 msec, and $N_b$ at 40–60 msec. The first or second peak may be identical to the slow brainstem BAEP. A sixth peak, $P_b$, at 50–90 msec, is often found in recordings including the first 100 msec after the stimulus and represents the early peak ($P_1$) of the late cortical AEP. A decrease of stimulus intensity reduces the amplitude and increases the latency of the MLAEP.

The MLAEP has a wide distribution with a maximum over the frontocentral areas. An origin of the MLAEP near the auditory cortex has been postulated, and is supported by direct recordings from the surgically exposed temporal cortex, although depth recordings have suggested that the earlier peaks $N_0$, $P_0$, and $N_a$ may be generated by subcortical structures.

A few maturational studies suggest that the latency of MLAEP peaks decreases only slightly between infancy and adulthood.

### 14.3.3 Neurologic Disorders That Cause Abnormal MLAEPs

MLAEPs have been used infrequently in the diagnosis of neurologic diseases. When recorded in combination with BAEPs, they have been reported to increase the incidence of abnormal findings in MS and acoustic neurinomas. Unilateral temporal lesions usually leave the $N_a$ and $P_a$ peaks of the MLAEP intact. Bilateral temporal lobe lesions have been found to abolish the MLAEP without abolishing the BAEP in cases showing cortical deafness but to leave the MLAEP intact in a case of audiologic agnosia.

### 14.3.4 Audiological Disorders That Cause MLAEPs

Audiological interest in the MLAEP is raised by reports suggesting that the MLAEP threshold lies within about 20 dB of the hearing threshold during wakefulness and sleep and that MLAEPs elicited by tone pips of different frequencies are specific for the tonal frequency of the stimulus even at relatively low frequencies. However, the usefulness of the MLAEP in the evaluation of hearing loss has not yet been convincingly proved. The threshold of the MLAEP may be as high as 60 dB HL in normal sleeping infants and MLAEPs may be absent in some normally hearing subjects, especially when tone pips rather than clicks are used. Because of this variable relationship between MLAEP and hearing threshold and because of the frequent contamination by muscle responses, the test is often useless for estimating hearing level, especially in young children where it is most needed. However, recording of the MLAEP in addition to the BAEP has been said to improve threshold evaluation. As a minimum, the MLAEP may be of value in a qualitative manner: Its presence, like that of a postauricular response, may be taken as evidence that conduction from the cochlea to the central nervous system, and therefore hearing, is not entirely absent.

## 14.4 THE 40-HZ AEP

Stimulation with clicks or tone pips that repeat at 40/sec and recording between electrodes on the forehead and the stimulated ear using a bandwidth of 10–100 Hz produce a steady-state AEP that is probably composed of MLAEP components (Figure 14.4). This 40-Hz AEP has a threshold of up to about 35 dB above hearing threshold during wakefulness. It can be studied with pips of a wide range of tonal frequencies including fairly low frequencies. The 40-Hz AEP may therefore become a useful hearing test for low tone frequencies.

## 14.5 THE LONG-LATENCY AEP

### 14.5.1 Methods of Stimulating and Recording

The subject should be alert and relaxed during the recording. Sleep, sleep deprivation, and changes of attention change the LLAEP. Infants and children often can not be examined while awake and must be sedated. Relaxation of neck

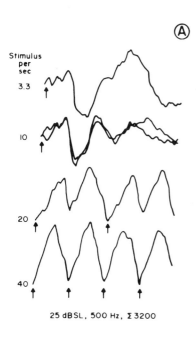
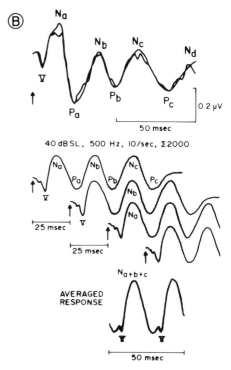

FIGURE 14.4. *The normal 40-Hz AEP. (A) Stimulation with weak tone bursts repeating at 3.3 and 10/sec produces MLAEP waves with peaks at 25 and 50 msec (top two tracings). Stimulation at 20 and 40/sec produces rhythmical steady-state AEPs with peaks at intervals of 25 msec, that is, at a frequency of 40/sec (bottom two tracings). (B) Intermittent stimulation with stronger tone bursts produces a complete transient MLAEP consisting of waves with a period of about 25 msec (top tracing). Repetition of the stimulus at intervals of 25 msec, or at a frequency of 40/sec, leads to superimposition of MLAEP waves, the presumed mechanism of the 40-Hz AEP (bottom tracings). Each stimulus consists of a 500-Hz tone burst of 6 msec at 25 dB SL in (A) and at 40 dB SL in (B). Recordings between forehead and stimulated ear. Negativity at the vertex is plotted upward. From Galambos et al. (PNAS 78:2643, 1981) with permission of the authors and New York Academy of Sciences.*

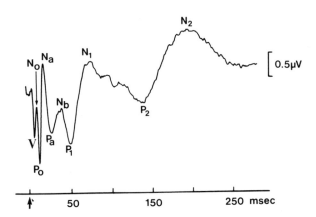

FIGURE 14.5.   *The normal LLAEP. This tracing is the grand average of LLAEPs from 35 normal subjects. The LLAEP waves P$_1$, N$_1$, P$_2$, and N$_2$ are preceded by BAEP and MLAEP waves. The LLAEP wave P$_1$ is identical with the MLAEP wave P$_b$. Binaural click stimulation at about 75 dB SL (arrow). Recording between vertex and mastoid. Negativity at the vertex is plotted upward. Slightly modified from Robinson and Rudge (Brain 100:19, 1977) with permission of the authors and Oxford University Press.*

and scalp muscles is not so critical because myogenic components are unlikely to obscure LLAEPs. Stimuli may be delivered from loudspeakers unless monaural testing is needed.

Stimuli usually consist of tone bursts of 250–2,000 Hz. To avoid generating a response to the onset of the tone burst, the burst is given a gradual rise over a time of 25–50 msec and a plateau of 30–50 msec. Stimuli are usually repeated regularly at 0.5–2/sec, even though larger responses may be obtained at slower rates and with irregular stimulation. Stimulation at rates faster than 2/sec may change the wave shape and latency. A LLAEP can also be elicited by changes in tone intensity, frequency, and apparent location in space.

Recordings are usually made in a single channel between vertex and one ear or mastoid. Displacement of the vertex electrode by up to 6 cm does not change the response. Amplification is not as high as for BAEPs or MLAEPs because the LLAEP amplitude is higher, ranging from 1 to 10 µV. Filter settings that include a bandwidth of 0.2–100 Hz are ample; narrower settings of only 2–15 Hz may be used. The sweep length is usually about 500 msec. Only about 30–100 responses need to be averaged for one LLAEP.

## 14.5.2 The Normal LLAEP

The LLAEP has an inconstant vertex-positive peak P$_1$ at 50–70 msec, a fairly large negative peak N$_1$ at 100–150 msec, and a positive peak

P$_2$ at 170–200 msec (Figure 14.5). The prominent N$_1$–P$_2$ complex is usually followed by a negative peak N$_2$. A third positive peak P$_3$ at about 300 msec, the P300, and peaks of even longer latency depend on cognitive processes rather than the stimulus. Long tone stimuli cause a sustained negative potential shift with a delayed onset.

The amplitude of the LLAEP decreases and its latency increases with increasing stimulus rates, with increasing stimulus rise time, and with decreasing stimulus duration. Similar changes occur with decreasing stimulus intensity, especially near threshold, and more so for click than for tone stimuli. Tones of low frequency elicit larger LLAEPs than high tones of equal sensation level, and the amplitude of LLAEPs to low tones increases more with increasing stimulus intensity than does the amplitude of LLAEPs to high tones. With increasing stimulus intensity, the LLAEP reaches a maximum after which a further increase of intensity may cause a decrease of amplitude. Latency does not depend on the tonal frequency of the stimulus.

The LLAEP has a wide distribution with a maximum at the vertex. The distribution varies with stimulus frequency. Responses recorded from the side opposite the stimulated ear are slightly larger. A local origin of the LLAEP in the sylvian area near the auditory cortex has been suggested, but not directly confirmed, by scalp recordings. Recordings made directly from the surface of the auditory cortex showed AEPs different from the LLAEPs obtained on the scalp. A positive peak at 105 msec followed by negativity at 150–160 msec has been isolated by scalp recordings from the temporal area and was thought to indicate excitation of the secondary auditory cortex. The reduction of $N_1$ by temporoparietal lesions suggested that this peak is generated by posterior-superior temporal and adjacent parietal areas. In general, it seems likely that the scalp-recorded LLAEP is produced by multiple generators, including the auditory cortex.

The effect of age on the LLAEP has been studied extensively. In premature infants, LLAEP peaks develop in keeping with EEG patterns. Even in newborns, the LLAEP is quite variable and depends on the sleep stage. The variability decreases with age. In general, the latency decreases and the amplitude increases, mainly during the first year of life. In adult and elderly subjects, LLAEP latency increases and amplitude decreases.

### 14.5.3 Neurologic Disorders That Cause Abnormal LLAEPs

MS was found to produce abnormal LLAEPs in a few patients, but LLAEPs were diagnostically less effective than BAEPs and MLAEPs.

Head injuries produced LLAEP changes that reflected severity and predicted outcome. Strokes and tumors in the temporoparietal regions, but not in the frontal regions, reduced the $N_1$. LLAEPs may be absent in postanoxic encephalopathy and in brain-damaged children. The latency of the LLAEP was found to be increased in Friedreich's ataxia.

LLAEPs of hyperkinetic children differ from those of normal children. Amplitude reductions have been described in drowsy narcoleptics. Schizophrenia and affective psychosis may produce nonspecific changes of LLAEP that tend to disappear with antipsychotic drug treatment.

Drugs such as alcohol, diazepam, amitriptyline, and dextroamphetamine may alter the LLAEP.

### 14.5.4 Audiologic Disorders That Cause Abnormal LLAEPs

Early studies raised hopes that the LLAEP may become a useful test of hearing at a wide range of frequencies. Although useful measurements may be obtained in healthy adults, the correlation between LLAEP and hearing deteriorates in patients with hearing loss. Large discrepancies between LLAEP threshold and audiometric threshold have been reported. The correlation between LLAEPs and hearing is worse in sleep for adults and even more so for young children. The LLAEP is only rarely useful in young children. LLAEP studies of hearing loss can therefore give only rather inexact results, but have been used to detect gross hearing losses in children with cerebral damage and mental retardation and to distinguish nonorganic, including hypnotically induced, hearing loss. Grossly, the presence of a LLAEP suggests that a subject can hear, but the absence of LLAEPs in clinical investigations can not be accepted as definite evidence of hearing loss. The BAEP, ECochG, and MLAEP may provide better audiometric results than the LLAEP, especially in cases where stimulus tone frequency is not essential.

## 14.6 THE ELECTROCOCHLEOGRAM

In the past, the ECochG has usually been recorded alone mainly for audiologic purposes. However, because the ECochG contains the equivalent of wave I of the BAEP, it has been suggested that the ECochG be recorded together with the BAEP in those cases in which a recording of wave I is essential but can not be obtained by BAEP recording methods. Many reviews of the ECochG have been published.[18,93]

### 14.6.1 Methods of Stimulating and Recording

Stimuli consist of tone pips or clicks delivered through earphones. Bone vibration may also be used when faulty air conduction is suspected. Masking the opposite ear is not necessary unless the BAEP is also recorded because the potentials comprising the ECochG can only be recorded from the stimulated ear.

A needle electrode penetrating the eardrum is more effective than electrodes in the lumen or the wall of the auditory canal for recording all three components of the ECochG. However, placement of such an electrode requires general anesthesia in children and local anesthesia in adults. The needle electrode is inserted by an otologist through the tympanic membrane so that its tip lies against the promontory close to the round window niche (transtympanic method). Atraumatic electrodes placed into the external auditory canal or needle electrodes inserted into the wall of the canal are more commonly used (extratympanic method). The reference electrode may be placed on the ipsilateral earlobe or mastoid. To obtain recordings that include the BAEP after the ECochG, the reference electrode is placed on the vertex. Single-channel recordings are made between the ECochG electrode and mastoid or ear reference. Two-channel recordings may add the combination of ECochG and vertex electrodes for a combined ECochG and BAEP recording. Earphones must have provisions to accommodate the recording electrode. The lead wire of the electrode must be shielded to reduce stimulus artifact. Amplifier gain is adjusted for the signal size, which may vary from less than 1 $\mu$V to about 10 $\mu$V. Bandpass should range from 10 to 3,000 Hz; narrower limits may be used to isolate the ECochG for analysis periods of about 5–10 msec.

The ECochG has three components: (1) the auditory nerve action potential (NAP), (2) the cochlear microphonic (CM), and (3) the summating potential (SP). To separate these three components, condensation and rarefaction stimuli are given alternately and the responses are averaged separately. The NAP, which always has the same polarity regardless of stimulus phase, can be separated from the CM, which changes phase with the stimulus, by adding the responses to both stimulus phases. This enhances the NAP in the average and reduces the CM like other noise (Figure 14.6). The NAP may be slightly distorted by this method because NAPs to condensation and rarefaction stimuli may have slightly different latency and shape, especially at low stimulus frequencies. The high-frequency filter should be set to at least 3 kHz to avoid distortion of the NAP. High-frequency filtering can therefore not be used effectively to

reduce the CM. The CM is isolated by subtracting the responses to stimulation with opposite phase from each other; this tends to cancel the NAP and to increase the CM (Figure 14.6). The CM may also be enhanced by filtering the responses with a narrow bandpass filter centered at the stimulus frequency. The CM may be distinguished from electric stimulus artifacts by introducing a piece of tubing between stimulus source and ear. This delays the sound stimulus at the ear against the electric artifact and attenuates the amplitude of the artifact; clamping the tube eliminates the sound stimulus and identifies the artifact. The SP is isolated either by using a high-frequency filter of less than 100 Hz to eliminate the NAP and CM or by stimulating at rates of 125–250/sec, which nearly abolished the NAP without affecting the SP.

### 14.6.2 The Acoustic Nerve Action Potential

The normal nerve action potential consists of a single peak, negative at the ear electrode, which has a latency of 1–2 msec in response to a tone pip or click of 60–90 dB above threshold (Figure 14.6). The NAP is the compound action potential generated by acoustic nerve fibers. Its latency increases with decreasing stimulus intensity and is 4–6 msec nearer threshold. The latency to stimuli containing high frequencies is shorter than that to stimuli of lower frequencies. The amplitude of the NAP decreases with decreasing tonal stimulus frequency; responses to 500 Hz and less are unreliable. Increasing the stimulus repetition rate also decreases NAP amplitude.

The NAP threshold is an excellent measure of hearing threshold for tone frequencies of over 1,000 Hz and is especially useful in children. It is fairly frequency specific but not very useful at low tone frequencies because it has a threshold of up to 60 dB above hearing threshold at 500 Hz. Specific combinations of threshold values and of relations between stimulus intensity and response latency and amplitude characterize different types of hearing loss. Conductive hearing loss has also been evaluated with bone vibration stimuli.

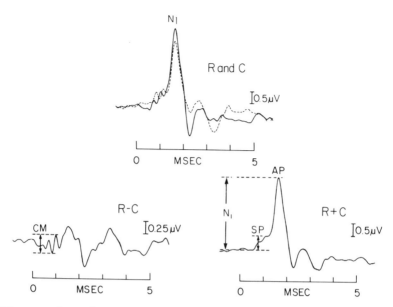

FIGURE 14.6. *The normal ECochG to click stimuli. Top: Averages of responses to rarefaction (solid line) and compression (dashed line) clicks, showing mainly the N₁ peak of the nerve action potential. Bottom left: Subtraction of responses to compression clicks from responses to rarefaction clicks enhances the cochlear microphonic (CM), which reverses phase with reversal of the stimulus polarity and cancels the action potential and the summating potential, which do not reverse phase with the stimulus polarity. Bottom right: Addition of responses to rarefaction and compression enhances both the N₁ peak of the action potential (AP) and the summating potential (SP) and cancels the CM. Stimulation with clicks of 115 dB peSPL. Recording between an electrode at the wall of the external ear canal near the tympanic membrane and an electrode on the bridge of the nose. Negativity at the ear electrode is plotted upward. From Coats (Arch Otolaryngol 107:199, 1981) by permission of the author and the American Medical Association.*

The NAP, in combination with the CM and SP, has also been used to evaluate sudden hearing loss, the effect of ototoxic drugs, and Ménière's disease. Even acoustic neurinomas and MS have occasionally been reported to produce NAP abnormalities. In general, however, the NAP is limited to the study of the most distal part of the auditory pathway. For investigations of hearing loss, recording of the BAEP is often preferred to recording of the NAP of the ECochG because it gives very similar and nearly equally reliable information with simpler methods. However, the BAEP recordings that do not show wave I may be greatly improved by simultaneous ECochG recordings, which have a much better chance to pick up auditory nerve potentials.

### 14.6.3 The Cochlear Microphonic

The cochlear microphonic (CM) consists of a series of rhythmical low-amplitude deflections that coincide with the peaks of the sound wave stimulus and reverse polarity with reversal of the stimulus phase (Figure 14.6). The exact shape of the CM depends on the location of the stimulus electrode. The CM, representing the phasic component of the reaction of the cochlear hair cells to sound waves, provides a crude measure of intactness of the cochlear sound receptors and may help in the qualitative evaluation of hearing loss. The CM may be used to divide sensorineural hearing losses into (1) sensory hearing loss characterized by absence of both the CM and NAP and (2) neural hearing loss having a preserved CM and an absent NAP. The CM may be preserved at low stimulus frequencies of about 500 Hz in cases of high-frequency hearing loss. The threshold of the CM does not correspond with hearing threshold, and the CM is therefore not useful in estimating the degree of hearing losses.

### 14.6.4 The Summating Potential

The summating potential (SP) consists of a small deflection of the baseline that has the same polarity as the NAP, which may be superimposed on the SP at low stimulus frequencies (Figure 14.6). The SP may represent a steady component of the receptor potential of cochlear hair cell. The SP can not indicate hearing but is abnormally increased in Ménière's disease and normalized by treatment with glycerol.

## 14.7 SONOMOTOR AEPS

### 14.7.1 Methods of Stimulating and Recording

The subject must not be allowed to relax because relaxation reduces scalp and neck muscle tone and may abolish the sonomotor AEP. Tension of neck muscles, intentionally produced by neck extension by resistance, enhances the sonomotor AEP. Effective stimuli have a fast rise time and a fairly high intensity of 50–80 dB HL; tones of 100 Hz and less are ineffective. Stimuli may be repeated at up to 10/sec but the responses may fatigue after two minutes or more.

Sonomotor AEPs may be picked up with electrodes placed near scalp or neck muscles. Postauricular AEPs may be recorded with an electrode up to 2 cm behind the ear with reference to an electrode on the earlobe. Other sonomotor AEPs are often effectively recorded with an electrode on the inion with reference to

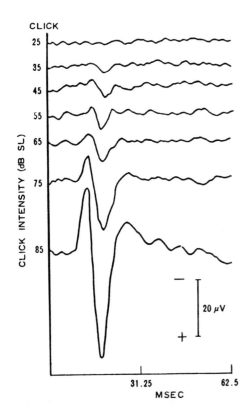

FIGURE 14.7. *Normal postauricular AEPs to stimulation with clicks of increasing intensity. Recording made with the subject sitting upright in a chair without headrest. Electrodes placed behind the stimulated ear and on the earlobe. Negativity behind the ear plotted upward. From Yoshie and Okudaira (Acta Oto-Laryngol Suppl 252:89, 1973) with permission of Actas Otolaryngologica.*

a distant electrode. The recording methods are otherwise similar to those used for MLAEPs.

## 14.7.2 The Normal Sonomotor AEP

Sonomotor AEPs are surface recordings of muscle contractions produced by sudden sound stimuli. As recorded with scalp or neck electrodes, normal sonomotor AEPs vary enormously depending on the recording site and the tone of underlying muscle. They may begin 6–50 msec after the stimulus and last up to 100 msec. In contrast to neurogenic AEPs, sonomotor AEPs have a maximum at the periphery of the scalp. Unilateral stimulation produces bilateral sonomotor AEPs. The postauricular AEP, also called *crossed acoustic response* because of its bilateral distribution, has a peak latency of 12–15 msec (Figure 14.7) but is also quite variable and depends on head position and filter settings. Unlike other sonomotor AEPs, the postauricular AEP may persist in sleep. Sonomotor responses may be recorded even in premature infants.

The postauricular response is probably mediated by excitation of the cochlea and therefore better suited for hearing tests than the response of neck muscles at the inion which is probably mediated by the labyrinth and may be preserved in deaf persons.

## 14.7.3 Neurologic Disorders That Cause Abnormal Sonomotor AEPs

The postauricular AEP has been used to study facial nerve lesions and to detect subclinical lesions in MS. However, it is not used routinely for either clinical postion.

## 14.7.4 Audiological Disorders That Cause Abnormal Sonomotor AEPs

Although attempts have been made to use the postauricular AEP for auditory threshold determinations, this AEP is generally considered to be too variable for quantitative audiometry. Because it may be absent in subjects with normal hearing, the postauricular AEP can be used only as a screening test, its presence suggesting that hearing is present.

# D

# Somatosensory Evoked Potentials

## PART CONTENTS

# 15

# SEP Types, Principles, and General Methods of Stimulating and Recording

## 15.1 SEP TYPES

Table 15.1 gives a breakdown of SEPs. They are distinguished mainly by the location of the stimulus and recording electrodes. In this regard, classification of SEPs differs from that of VEPs and AEPs, although it resembles AEP classification in that it depends more on response latency than on stimulus type. Most SEPs are produced by electric shocks applied to the nerves of the arm or leg. SEPs to arm nerve stimulation are usually elicited by stimulating the median nerve at the wrist. They are recorded stimultaneously with electrodes placed at Erb's point above the clavicle (Erb's point potential, EP), the neck (cervical SEP), and the parietal scalp (scalp SEP), reflecting activity generated mainly in the brachial plexus, the upper cervical cord, and the somatosensory cortex, respectively. In a different approach, the effect of arm nerve stimulation may be recorded by far-field techniques using two widely spaced electrodes to record a single SEP that consists of a sequence of peaks representing excitation of the brachial plexus,

spinal cord, and somatosensory cortex. SEPs to stimulation of arm nerves may be further characterized by the nerve stimulated: Stimulation of different peripheral nerves produces SEPs of slightly different latency and distribution.

SEPs to leg stimulation are usually elicited by stimulating the posterior tibial nerve at the ankle or the common peroneal nerve at the knee. They are recorded over the lumbar (lumbar SEP) and lower thoracic (thoracic SEP) spine and the scalp (scalp SEP), reflecting activity of the cauda equina, lower spinal cord, and somatosensory cortex, respectively. Far-field recordings may show peaks of subcortical origin preceding the cortical SEP. Recordings from the upper thoracic and the cervical spine usually do not show reliable SEPs to leg stimulation. In the case of posterior tibial nerve stimulation, recording over the popliteal fossa may be added to better evaluate peripheral sensory conduction. Sensory nerve action potentials may be studied by simulating a purely sensory nerve and recording from the same nerve proximal to the stimulus point. SEPs not routinely used in clinical practice include SEPs to stimulation of the trigeminal nerve, skin

TABLE 15.1. SEP types

A. SEPs to electric stimulation of arm nerves
    1. Clavicular SEP (Erb's point potential)
    2. Cervical SEP
    3. Scalp SEP
    4. Far-field SEP
B. SEPs to electric stimulation of leg nerves
    1. Lumbar SEP
    2. Low thoracic SEP
    3. Scalp SEP
    4. Far-field SEP
C. Sensory nerve action potential (SNAP)
D. SEP to electric stimulation of the trigeminal nerve
E. SEP to electric stimulation of pudendal and bladder mucosa
F. Steady-state SEP to repetitive electric stimulation
G. SEPs to touch, joint position change, and other stimuli
H. Somatomotor SEP

of dermatomes, pudendal and bladder mucosa, SEPs to stimuli other than electric shocks, and steady-state SEPs of any kind.

Most, if not all, types of SEPs are mediated by fibers in the dorsal columns of the spinal cord, the medial lemniscus system of the brainstem, and the nucleus ventralis posterolateralis and posteromedialis of the thalamus. This is suggested by the frequent clinical observation that SEPs are rendered abnormal by lesions that cause loss of vibration and position sense rather than by lesions that reduce pain and temperature sensation. Muscle afferents traveling with cutaneous afferents in the dorsal columns probably contribute to the shortest latency peaks elicited by stimulation of mixed motor and sensory nerves. Fibers that travel in the anterolateral part of the spinal cord and extralemniscal systems of the brainstem and mediate pain and temperature sensation may contribute to some cortical peaks of long latency.

# 15.2 THE SUBJECT

The subject should lie comfortably on a bed or in a reclining chair. Muscles at the recording sites, especially scalp and neck muscles, should be relaxed to avoid contamination of SEPs by muscle potentials. If muscle artifact can not be eliminated by comfortable positioning of the subject and by other attempts at relaxation, a sedative or mild hypnotic may be used. The subcortical and short-latency cortical SEPs recorded in routine clinical studies are not significantly altered by changes of attention, sleep, or CNS-depressant drugs. These factors generally affect only peaks of longer latency. Changes of attention may alter cortical SEP peaks occurring no sooner than 30 msec after arm stimulation, even though operant conditioning has been reported to be capable of altering even peaks of shorter latency. Sleep may change the amplitude, but not the latency of early cortical peaks; both the amplitude and latency of late cortical peaks are affected by sleep, especially in newborns. Alcohol and CNS-depressant drugs have been reported to alter scalp SEPs, although central conduction time was found to be independent of phenobarbital at serum levels of up to 146 µg/ml and of thiopental at levels suppressing the spontaneous EEG. Phenytoin at toxic serum levels has been reported to increase central conduction time. Active or passive movement of a stimulated finger may decrease the amplitude of cortical peaks appearing about 50 msec after the stimulus.

The limbs should not be allowed to become cool because decreases of temperature prolong peak latencies. Deep central hyperthermia and hyperthermia of 42°C may alter the SEPs. The distances between stimulating and recording electrodes should be measured to evaluate peripheral and central conduction velocities.

# 15.3 STIMULATION METHODS

## 15.3.1 Stimulus Types

### 15.3.1.1 Electric shocks

Electric shocks to a peripheral nerve are clinically the most useful stimulus type. The shocks may be delivered through surface or needle electrodes.

### 15.3.1.1.1 Types and application of stimulus electrodes. Surface stimulus electrodes consist

of EEG disc electrodes, bipolar EMG stimulating electrodes imbedded in a plastic strip, or ring electrodes for finger stimulation. The skin at contact points over a nerve is slightly abraded. Disc electrodes are glued into place with collodion and filled with conductive jelly. Attachment with sticky conductive paste is less secure. EMG stimulating electrodes may be held in place with adhesive tape or Velcro strip. Ring electrodes are moistened with conductive jelly and placed on individual fingers for stimulation of the digital nerves. Electrode impedance should be less than 10 kohms to reduce discomfort and stimulus artifact.

Needle electrodes consist of sharp wires of stainless steel or other inert metal, less than 1 mm in diameter and about 1 cm long. They are inserted perpendicularly through the cleansed and disinfected skin so that the electrode tips lie close to the nerve. The lead wires should be taped to the skin to keep the needles from slipping out. Although insertion causes brief pain, needle electrodes are generally well tolerated and have the advantage of generating less stimulus artifact in the recording because they require much less stimulus current than do surface electrodes. Because of the danger of infection, needle electrodes can not be used during long procedures such as surgical monitoring.

For nerve stimulation, the two stimulus electrodes of a pair are placed along the course of a nerve. Surface electrodes are separated from each other by 2–3 cm, needle electrodes by only about 1 cm. For dermatomal stimulation, both electrodes are placed a few centimeters apart along the middle of a dermatome. For both nerve and skin stimulation, the proximal stimulus electrode is connected to the negative pole of the stimulator output and becomes the cathode; the distal electrode is connected to the positive pole and becomes the anode. The two output terminals of the stimulator are isolated from ground through a radiofrequency or optoelectric isolation unit to avoid current flow from either stimulus electrode to the ground electrode, thereby restricting the stimulus current flow to the area between the two stimulus electrodes. Spread of stimulus current could increase the stimulus artifact and be a hazard to the subject. The subject's safety also requires that high stimulus intensity, long stimulus pulses, and repetitive stimulation at high rates be avoided, especially in patients who can not feel the stimulus.

*15.3.1.1.2 Stimulus electrode placements.* Arm stimulation commonly uses the median nerve at the wrist. Leg stimulation is usually applied either to the posterior tibial nerve at the ankle or to the common peroneal nerve at the knee. Because stimulation of these nerves yields relatively large and fairly constant SEPs, it is the best method for most clinical studies. In principle, however, stimulation of mixed motor and sensory nerves is less than ideal because it excites (1) heterogenous afferents from the skin, joints, muscle, and deep tissues producing orthodromically conducted SEP components in various pathways up to the level of the scalp, and (2) motor fibers generating antidromically conducted SEP components at the clavicular and lumbar levels. Some of the SEP components may persist with normal latency in cases of sensory neuropathies in which the SEP components generated by somatosensory input may be delayed or abolished; this may result in combinations of SEPs that are difficult to interpret. Therefore, in cases where testing of purely somatosensory input is essential, stimuli must be applied to sensory nerves or nerve branches. The precise electrode placement for stimulation of mixed and sensory nerves is described in chapter 16 for arm and leg nerves.

Stimulation of sensory nerves or nerve branches may also be used to detect the location of peripheral lesions, or the level of spinal cord lesions. This requires investigation of several inputs, which may tax the patience of the patient. Furthermore, SEPs to stimulation of small nerves differ from each other in latency and distribution on the scalp and require establishment of normal SEP values for each stimulus location. Also, they often have very low amplitude.

*15.3.1.1.3 Stimulus intensity.* For stimulation of mixed nerves, the intensity is usually set slightly above the motor threshold, that is, at an intensity sufficient to produce a muscle contraction causing a visible twitch. If no twitch can be elicited, as in severe peripheral neuropathy, the stimulus should have an intensity known to produce a visible twitch in normal subjects. This usually requires stimulus currents of 5–15 mA for stimulation with surface electrodes and weaker currents for stimulation with needle electrodes.

For sensory nerve stimulation, the stimulus intensity is set at 2.5–3 times the sensory threshold. When stimulating an anesthetic area, the stimulus intensity for purely sensory nerves may be set by the effect on simultaneously recorded sensory nerve action potentials. Especially in cases of superficial sensory loss and of unconscious patients, stimulus intensity must be kept moderate to avoid local damage.

The definition of stimulus intensity by motor and sensory thresholds is more reliable than the use of stimulus voltage or current. Stimulus voltage is not a precise measure of stimulus intensity because the effect of electric stimulation depends on current flow: The current generated by a stimulus voltage depends on the impedance of the stimulus electrodes and of the tissue between them. Even stimulus current is not an entirely satisfactory measure because the fraction of current crossing the nerve fiber membranes and acting as a stimulus varies with the distance between, and the orientation of, stimulating electrodes and nerve. However, in situations where constant stimulus electrode impedance is difficult to maintain, such as during surgical monitoring, a stimulator capable of delivering constant current output can compensate for changes of stimulus electrode impedance and thereby help to maintain constant stimulus conditions.

*15.3.1.1.4 Stimulus duration.* The duration of the electric pulse is usually 200 μsec, especially for stimulation of mixed nerves. Shorter stimuli tend to excite a larger proportion of motor than of sensory fibers.

### 15.3.1.2 Other sensory stimuli

Touch, vibration, joint movement, muscle stretch, pain, and temperature changes are rarely used for clinical SEP studies.

### 15.3.2 Stimulus Rate

Transient SEPs are usually elicited with stimulus rates of 4–7/sec. Rates up to 10/sec do not alter the clinically important short-latency peaks, including early scalp SEP peaks, but may be uncomfortable for the patient. Slower rates must be used for studies of the clinically less important late peaks of the scalp SEP. Steady-state cortical SEPs may be elicited at rates of 12–200/sec but have been used only rarely. The stimulus rate should not be an integral of 60 Hz to avoid buildup of power-line artifact.

### 15.3.3 Unilateral and Bilateral Stimulation

For arm stimulation, only one side of the body is usually stimulated at a time to aid proper lateralization of abnormalities. Simultaneous bilateral stimulation may give better defined SEPs and, in conjunction with bilateral scalp recording, has been said to detect hemispheric lesions not detected by unilateral stimulation of either side alone. However, bilateral stimulation can obscure unilateral abnormalities and does not permit lateralization of abnormalities in subcortical structures. Routine studies therefore use unilateral stimulation. The same is generally true for leg stimulation except where lateralization of a lesion is not critical, for instance, during surgical monitoring of spinal cord function. Here, nerves of both legs may be stimulated simultaneously to obtain larger scalp SEPs. However, simultaneous bilateral stimulation is acceptable only if peripheral conduction distances and velocities are equal because a delay of one peripheral afferent can destroy the scalp SEPs to combined stimulation and suggest central problems.

## 15.4 GENERAL RECORDING PARAMETERS

### 15.4.1 Recording Electrode Placements

The volley of impulses ascending in the stimulated somatosensory pathway is usually recorded with electrode pairs placed at different levels of that pathway, one electrode of each pair being located as closely as possible to the structures generating the SEP peak at that level. On the other hand, widely spaced electrodes may be used to record a far-field SEP containing several peaks generated by the various relays of the somatosensory pathway. (Specific recording electrode placements and combinations are described in the sections on arm and leg SEPs.) The subject is grounded through a large metal plate or band covered with electrode jelly or saline-soaked gauze and placed between the stimulating and recording electrodes to minimize the stimulus artifact. The ground electrode is connected to the ground input of the recording amplifier of the groundpost of the stimulator so that the subject, the recording equipment, and the stimulator all lie at the same ground potential level. The ground electrode should have an impedance of less than 10 kohms. As in all EP recordings, a ground electrode is needed to protect the subject and to reduce the chance for pickup of interference artifacts. In SEP recordings in particular, proper grounding is necessary to keep the stimulus artifact to a minimum. If the stimulus artifact is excessive, some relief may be obtained by reducing the impedance of the ground electrode or by changing its connection from amplifier input to stimulator groundpost or vice versa. Other measures apt to reduce the stimulus artifact include reducing the impedance of the stimulus electrodes, changing the orientation of stimulus electrodes to recording electrodes, and changing the filter settings, especially of the low-frequency filter.

### 15.4.2 Filter Settings

A wide bandpass from 5–30 Hz to 2,500–4,000 Hz is usually used for all channels. Reduction of the high-frequency filter can substantially increase the latency. If filters are set differently for each channel, the high frequencies could be cut more for cortical SEPs and the low frequencies could be cut more for the other SEPs. Filter roll-off slopes should not exceed 12 dB/octave for low frequencies and 24 dB/octave for high frequencies.

### 15.4.3 Other Recording Parameters

Averagers with four channels are commonly used to record SEPs from different locations simultaneously. Different sweep lengths and numbers of sweeps are used for arm and leg SEPs. At least two sets of averages must be superimposed to ascertain replication.

# 16

# Normal SEPs to Arm Stimulation

## 16.1 NORMAL SEPS AT DIFFERENT RECORDING SITES

A great variety of stimulating and recording methods are used in different laboratories. Older studies often recorded from only one or two sites. Because averagers with four channels have now become widely available, current practice favors simultaneous recordings from three or four points along the somatosensory pathway. A common technique uses stimulation of the median nerve at the wrist and recording from clavicular, cervical, and scalp electrodes (Figure 16.1). Other methods use different stimulus and recording sites. SEPs to arm stimulation vary depending on subject variable, stimulus characteristics, and recording parameters. Commonly used conditions are summarized in Table 16.1

### 16.1.1 The Clavicular SEP (Erb's Point Potential)

Clavicular SEPs may be recorded from Erb's point 2–3 cm above the clavicle at the posterior border of the clavicular head of the sternocleidomastoid muscle, which can be easily seen when the subject bends his head against a resistance. Electrode placements more lateral over the clavicle may give recordings of higher amplitude and only slightly different latency. The clavicular electrode on the stimulated side may be referred to the opposite clavicular electrode. If the clavicular electrode on the stimulated side is referred to a midfrontal electrode, that is, Fz in the International 10–20 System of EEG electrode placements, the recording may contain later peaks that reflect activity of the central parts of the somatosensory pathway and follow the early peak picked up by the clavicular electrode.

Clavicular SEPs have a major negative peak, often preceded and followed by a smaller positive wave; smaller negative waves may follow this first complex (Figure 16.2). Latency is measured to the peak, or less commonly, to the onset of the negative wave. In normal adults, peak latency amounts to about 9 msec for stimulation of the median nerve at the wrist ($N\overline{9}$) and about 11 or 12 msec for digital nerve stimulation, depending on age and arm length. The negative

171

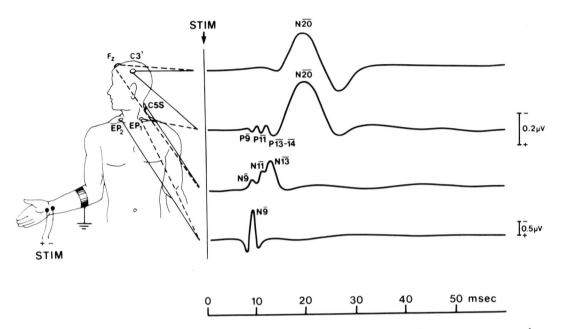

FIGURE 16.1. *Schematic diagram of normal SEPs to arm stimulation. Tracings, from bottom to top, show clavicular SEP (Erb's point potential, N$\overline{9}$), cervical SEP (N$\overline{13}$), far-field SEP recorded between scalp and noncephalic reference electrodes, and scalp SEP recorded between scalp and cephalic reference electrodes (N$\overline{20}$). Negativity at the electrodes connected with a solid line in the diagram at the left is plotted upward.*

peak is the benchmark from which central conduction times to cervical and scalp SEP peaks are measured.

The clavicular SEP is generated by the peripheral nerve fibers contained in the brachial plexus. In case of sensory nerve stimulation, the clavicular SEP is generated by somatosensory fibers only. When a mixed nerve is stimulated, orthodromic responses of muscle afferents and an antidromic motor volley are added to the somatosensory component.

## 16.1.2 The Cervical SEP

Cervical SEPs may be recorded from the neck at the level of the C5 or C2 vertebra, that is, two or five spinous processes above C7 which has the most prominent spinous process at the base of the neck; C2 lies at the point of the deepest indentation a few centimeters below the inion. In some laboratories, cervical SEPs are recorded from an electrode on C7. Recordings from neck

electrodes commonly use a midfrontal reference electrode.

SEPs recorded with a cervical spine-scalp derivation have, as their largest and most consistent part, a negative peak with a latency at about 13 msec (N$\overline{13}$) after stimulation of the median nerve at the wrist (Figures 16.1 and 16.2). This peak may be preceded and followed by smaller negative peaks at 9 msec, 11 msec, and 14 msec (N$\overline{9}$, N$\overline{11}$, and N$\overline{14}$). The latencies are longer after stimulation of the digital nerves. Latencies and conduction times vary among laboratories, mostly depending on the location of the stimulating and recording electrodes.

In general, the generators of these peaks have not yet been clearly identified. However, it is now commonly accepted that the N$\overline{9}$ peak is due to excitation of peripheral nerve fibers between axilla and spinal cord. N$\overline{11}$ is probably generated at the level of the cervical cord segments by the dorsal root entry zone, gray matter of the dorsal horn, and nearby dorsal columns;

TABLE 16.1. SEPs to stimulation of the median nerve at the wrist

A. Subject variables
  1. Age: Several separate normal control groups are needed for children up to about 8 years; requirements for old age have not yet been defined.
  2. Sex: Separate control groups for adult males and females may be used as needed in each laboratory.
  3. Limb length: Distance from stimulus to recording electrode must be measured, especially for determination of peripheral conduction velocity.
  4. Temperature: To avoid cooling, the room temperature should be kept at 22–24°C.
  5. Sensory disturbances: Loss of senation must be defined in terms of distribution, modality, and degree. Stimulus intensity must be adjusted in anesthetic areas.
B. Stimulus
  1. Type: Electric shocks from electrically isolated constant voltage or constant current stimulator
  2. Electrodes: EEG disc electrodes of less than 10 kΩ impedance or needle electrodes
  3. Electrode placement: Cathode over median nerve at the wrist; anode 2–3 cm distal or on dorsum of wrist
  4. Rate: 4–7/sec
  5. Duration: 200–300 μsec
  6. Intensity: Above threshold for thumb abduction twitch
  7. Unilateral stimulation: Each side should be stimulated separately in routine studies.
C. Recording
  1. Number of channels: 4
  2. Electrode placements:
    a. Left and right Erb's point (EP1, EP2)
    b. Over C2 or C5 spinous processes on the neck (C5S, C2S)
    c. On the scalp 2 cm posterior to C3 and C4 positions (C3', C4')
    d. Midfrontal scalp (Fz)
  3. Montage for stimulation of the left median nerve (opposite recording side of lateral electrodes for stimulation of the right median nerve):
    a. Channel 1: C4'–Fz
    b. Channel 2: C4'–EP2
    c. Channel 3: C5S or C2S–Fz
    d. Channel 4: EP1–EP2
  4. Filter settings:
    a. Low-frequency filter: 5–30 Hz
    b. High-frequency filter: 2,500–4,000 Hz
  5. Number of responses averaged: 500–2,000
  6. Sweep length: 40 msec; 60–100 msec for delayed SEPs
D. Analysis
  1. Normal peaks:
    a. Erb's point potential ($\overline{N9}$) in channel 4
    b. $\overline{N13}$ in channel 3
    c. $\overline{P9}$, $\overline{P13/14}$, and $\overline{N20}$ in channel 2
  2. Criteria for abnormal
    a. Absence of normal peaks
    b. Slow peripheral conduction from stimulating electrode to Erb's point
    c. Increased central conduction times between Erb's point potential and cervical SEP ($\overline{N9}$–$\overline{N13}$), between cervical and scalp SEP ($\overline{N13}$–$\overline{N20}$), between brainstem far-field and scalp SEP ($\overline{P13/14}$–$\overline{N20}$), and between Erb's point potential and scalp SEP ($\overline{N9}$–$\overline{N20}$)
    d. Questionable criteria: Abnormal latency differences to stimulation of each side of the body; abnormal amplitude differences to stimulation of each side; abnormal amplitude ratios between central and peripheral SEPs

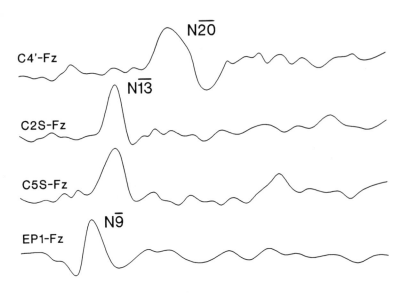

C4'-Fz

N20

N13

C2S-Fz

C5S-Fz

N9

EP1-Fz

FIGURE 16.2.　*Normal SEP to median nerve stimulation. Stimulation of the left median nerve at the wrist produces clavicular (bottom tracing), cervical (middle two tracings), and scalp (top tracing) SEPs. Recording electrodes are on Erb's point (bottom) over the spinous processes of C5 and C2 (third and second tracing, respectively) and on the scalp over the hand area on the right side. Negativity at these electrodes is plotted upward. Midfrontal reference electrode. The sweep is delayed by 4 msec to eliminate the stimulus artifact.*

N$\overline{13}$ and N$\overline{14}$ probably have several sources in the spinal cord, including the dorsal columns and nucleus cuneatus of the dorsal columns. The brainstem makes little or no contribution to these peaks. Direct recordings from electrodes near the cord indicate that at least part of the N$\overline{13}$ and N$\overline{14}$ waves are generated locally by a stationary source, probably the dorsal horn of the spinal cord or the dorsal column nuclei. Recordings from surface electrodes, showing a rostral increase of latency of some components, suggest that another part of the N$\overline{13}$ and N$\overline{14}$ peaks may be due to impulses ascending in the medial lemniscus, although lemniscal and higher parts of the somatosensory pathway are more clearly reflected in the far-field peak of P$\overline{13/14}$ recorded with widely spaced electrodes.

### 16.1.3 The Scalp SEP

Scalp SEPs to arm stimulation are recorded with electrodes over the contralateral parietal scalp, often placed 2 cm behind and C3 or C4 electrode positions of the International 10–20 System of EEG electrode placement (C3', C4'). These electrodes record a relatively large SEP to stimulation of the opposite upper limb. The early peaks vary depending on the location of the reference electrode.

#### 16.1.3.1 The scalp SEP recorded between parietal and midfrontal electrodes

The near-field scalp SEP to stimulation of the median nerve at the contralateral wrist consists of a series of waves. The clinically most important negative peak has a latency of about 20 msec (Figures 16.1 and 16.2). The N$\overline{20}$ may be preceded by earlier peaks that have a very low amplitude in recordings between parietal and midfrontal electrodes and can be better analyzed with widely spaced electrodes. The N$\overline{20}$ is followed by a positive peak at about 30 msec. This N$\overline{20}$–P$\overline{30}$ sequence is the first part of a W-shaped complex that is recordable in only about one half of all young normal adults; the second part of the W consists of a negative peak at 30–40 msec and a positive peak at about 40–50 msec. This complex may be followed by a slower negative wave with a peak latency at about 50–70 msec and a positive peak at about 80–90 msec. Later waves include a negative-positive sequence at 140–170 msec and 190–270 msec and a positive peak at about 300 msec. The latencies of these peaks increase from the front of the head to the back and are 2–3 msec longer on stimulation of the digital nerves than on stimulation of the median nerve at the wrist.

While later peaks are most likely generated by cortical relays, the generator of the N20 is still debated. An origin at the parietal cortex is supported by the limited distribution of the N20 in the parietal area contralateral to the stimulated arm, by the effect of cortical lesions, and by recordings made directly from the cortex. However, it has also been claimed that the N20 is generated in the thalamocortical radiation or the nucleus ventralis lateralis of the thalamus, although this does not seem to be supported by direct recordings from the thalamus or by the effect of thalamic lesions. Multiple cortical and subcortical generators have also been proposed.

SEPs recorded over the frontal scalp or cortex may show much independence of amplitude and latency from SEPs recorded simultaneously in the parietal areas, suggesting that frontal and parietal SEPs are generated separately, perhaps via independent cortical input from the thalamus.

Small SEP peaks of short latency can be recorded from the scalp ipsilateral to the stimulated arm and are probably conducted electrotonically from the contralateral hemisphere. Later peaks with latencies of over 40–60 msec may be due to activation of ipsilateral cortical areas through sensory input from bilateral pathways or through callosal connections between the hemispheres. The late N140–P190 complex and the P300 have a maximum at the vertex or posterior and lateral to it. Cerebral SEPs may be

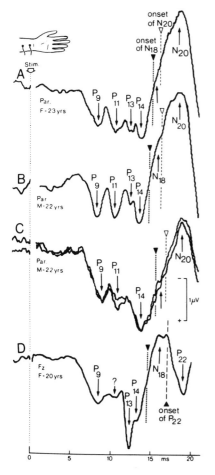

FIGURE 16.3. *Normal far-field SEPs to arm nerve stimulation in four young adults (A–D). Stimulation of the left median nerve at the wrist and recording between the contralateral parietal (A–C) or midfrontal (D) electrode and an electrode on the right hand produces a sequence of positive peaks (P9, P11, and P14) followed by a larger negative peak (N20). Two tracings are superimposed in (C). Negativity at the scalp electrodes is plotted upward. From Desmedt and Cheron (EEG Clin Neurophysiol 52:533, 1981) with permission of the authors and Elsevier Scientific Publishers Ireland Ltd.*

distorted by myogenic potentials that are triggered by the stimulus, begin at latencies under 20 msec, and last over 100 msec.

### 16.1.3.2 The scalp SEP recorded between parietal and ear or other distant cephalic electrodes

Recordings between the parietal electrode and electrodes on the ear or other scalp areas may pick up small peaks before N20 that are generated by subcortical structures. An early positive wave occurs 13–15 msec after stimulation of the median nerve at the wrist (P13/14) and has a maximum contralateral to the stimulated side. This wave probably originates from lemniscal input to the thalamus with possible contributions by the nucleus ventralis posterolateralis thalami and its thalamocortical outflow. This origin is supported by recordings from patients with lesions at different levels of the sensory pathway and by direct recordings from these structures in humans. Negative waves at 16 and 17 msec have also been thought to be generated by the thalamus or its radiation. Even earlier peaks, reflecting the activity of spinal cord structures usually recorded with cervical electrodes, may be picked up between parietal and various scalp reference electrodes. Recordings between parietal and ear electrodes may distort later peaks because the ear is not electrically indifferent since it picks up some SEP components.

### 16.1.3.3 The scalp SEP recorded between scalp and noncephalic reference electrodes

Recordings between a scalp electrode and an extracranial reference electrode may show up to six peaks: P9, P11, P13, P14, N20, and P27 (Figure 16.3). The three earlier ones are far-field SEPs that correspond with the negative peaks recorded at the neck. The P14 corresponds with the P13/14 recorded with widely spaced scalp electrodes. The N20 and P27 are probably near-field recordings of cortical potentials picked up mainly by the scalp electrode. A widely distributed N18 may be recorded before the N20 and is probably generated below the cortex. Most of the earlier peaks can be more unequivocally analyzed with clavicular and cervical record-

ings. The latency of the P9 in far-field recordings varies with shoulder position.

## 16.2 SUBJECT VARIABLES

### 16.2.1 Age

#### 16.2.1.1 From premature to adult age

Scalp SEPs of premature babies of over 24 weeks conceptional age show a slow, large negative wave of over 200 msec latency. With increasing age, the amplitude of this wave decreases while earlier peaks emerge. The scalp SEP appears first over primary sensory cortex and later over frontal association areas. Full-term newborns have SEPs that vary more with sleep stages than do those of adults, but often show an N30 on stimulation of arm nerves and a P50 after stimulation of leg nerves; each SEP is narrowly restricted to the cortical representation of the stimulated extremity. The SEP gradually reaches adult form and latency at an age between 3 years and 10 years. However, the latencies of SEPs of children can not be equated with those of adults: Because children have smaller bodies, latencies of the same length as in adults reflect much slower conduction velocities. Central conduction velocity increases from about 10 m/sec at birth and reaches adult values of about 50 m/sec at the age of about 8 years. In contrast, peripheral conduction velocity matures faster, from 20–25 m/sec at birth to the adult range of 60–80 m/sec at the age of 12–18 months. Changes of latency between ipsilateral and contralateral SEP peaks have suggested maturation of callosal connections.

#### 16.2.1.2 From adult to old age

Aging seems to affect SEPs less than other EPs. In octogenarians, all cortical peaks have slightly longer latency due to slowing of peripheral conduction velocity which was found to decrease from 71 m/sec at the age of 22 years to 61 m/sec at the age of 82 years in one study.[25] Another study showed peripheral conduction velocity to decrease throughout adult life at an annual rate of 0.16 m/sec.[27] Central conduction between cervical and scalp electrodes slows only slightly, if at all. The amplitude of the cervical SEP decreases progressively after the age of 40 years. The amplitude of the parietal SEP and the clear definition of the W shape increase with age. Age-dependent changes in the precentrally recorded SEP differ from those in the parietal area. Differential aging of separate parts of the somatosensory cortex has been suggested by the finding of different latency increases of the $\overline{N20}$ and the subsequent positive peak.

### 16.2.2 Sex

Several studies report that SEPs have longer latencies in men than in women; however, in one study the latency differences disappeared when differences in height were taken into account. Most laboratories do not use separate normal controls for the two sexes.

### 16.2.3 Limb Length

Absolute latencies depend on the distance between stimulus and recording electrodes which therefore must be measured for calculation of conduction velocities.

### 16.2.4 Limb Temperature

The stimulated limb must be kept warm by maintaining the room temperature at 20–22°C, by covering the limb, or by applying a heating pad as needed.

### 16.2.5 Sensation

Numbness is often present in patients with abnormal SEPs and requires careful adjustment of stimulus intensity.

## 16.3 STIMULUS CHARACTERISTICS

### 16.3.1 Stimulus Electrode Placement

Stimulus electrodes are most often placed on the median nerve at the wrist. One electrode is positioned 2 cm proximal to the wrist crease between the tendons of the palmaris longus and the flexor carpi radialis muscles, which can be easily seen when the subject flexes the wrist against resistance. This electrode is connected to the negative pole of the stimulator and thereby becomes the cathode. Another electrode is placed 2–3 cm distal to the cathode or on the dorsal surface of the wrist and is connected to the positive pole of the stimulator, thereby becoming an anode. The ground electrode consists of a plate on the palmar surface of the forearm or a band electrode around the forearm (Figure 16.1). Stimulation of the ulnar and radial nerves at the wrist has been used to detect lesions of other parts of the sensory input from the upper extremity. For stimulation of purely sensory nerves, stimuli may be applied:

- to median nerve fibers through ring electrodes placed over the digital nerves distal to the interphalangeal joint of the second or third finger or of both these fingers,
- to ulnar nerve fibers through ring electrodes over digital nerves of the fifth finger,
- to the superficial branch of the radial nerve at the radial side of the dorsal part of the wrist, and
- to the cutaneous branch of the musculocutaneous nerve two finger breadths below the lateral part of the cubital crease.

The cervical segments C5–C8 can be tested by stimulating various nerves or nerve branches at different points: C5 by stimulating the musculocutaneous nerve on the radial side of the forearm, C6 by stimulating the skin of the thumb, C7 by stimulating the adjoining surfaces of the second and third fingers, C8 by stimulating the skin of the fifth finger. C5 and C6 have also been studied by stimulating the musculocutaneous nerve near the wrist. Other segments have been studied by dermatomal stimulation.

### 16.3.2 Stimulus Intensity, Rate, and Duration

In the case of mixed nerve stimulation, the intensity is set slightly above the twitch threshold for a visible thumb twitch for the median nerve. In most laboratories, this requires an intensity of at least several milliamperes for surface electrodes, or of a few milliamperes for needle electrodes. In general, stimulating with needle electrodes should be avoided because of the risk of thermal damage due to high current density at the tip. The threshold for sensory nerves is determined as described earlier.

A change of stimulus intensity produces little or no significant change of the clinically important short-latency SEPs except that the amplitude decreases at low intensities. In general, peaks of longer latency have slightly higher thresholds than peaks of shorter latency. Stimulus rate is 4–5/sec. Stimulus duration is 200–300 μsec.

## 16.4 RECORDING PARAMETERS

Most recording parameters, including electrode placements, have been described and summarized in Table 16.1. About 500–2,000 responses must be collected for a SEP to stimulation of the median nerve at the wrist. Larger numbers are needed for the smaller SEPs produced by stimulation of smaller nerves. The sweep length is about 40 msec for most SEPs to arm stimulation. Longer sweeps may be required for SEPs of abnormally long latencies. (The importance of the sweep length for the wave frequencies in the SEP is discussed in chapter 18.)

## 16.5 GENERAL STRATEGY OF STIMULATING AND RECORDING

The strategy of testing the various parts of the somatosensory pathway from the arm to the cortex is straightforward for either of the two recording methods used. The more common method of recording SEPs simultaneously from clavicular, cervical, and scalp electrodes yields the peaks N9, N13, and N20 which indicate excitation of the brachial plexus, the upper spinal cord, and the cerebral cortex or its thalamic afferents, respectively. The less common method of recording between scalp and noncephalic reference electrodes reflects activity of the same structures in the successive peaks of a single SEP.

With both methods, a lesion between these structures either abolishes or delays and diminishes the peaks representing structures at the level of the lesions that involve the three segments of the somatosensory pathway between stimulation point and brachial plexus, between brachial plexus and upper cervical cord, and between upper cervical cord and somatosensory cortex or thalamocortical afferents.

- Peripheral nerve and plexus lesions delay or abolish all SEP peaks. The delay of the peak representing Erb's point is best expressed as a decrease in the peripheral nerve conduction velocity.
- Lesions of the cervical roots and cervical cord leave intact the clavicular SEP but delay or abolish the cervical SEP and, in most instances, also the scalp SEP. The delay causes an increase of the clavicular-cervical (N9–N13) and clavicular-scalp (N13–N20) conduction times in separate recordings from these points. Recordings between scalp and noncephalic electrodes show increased separation between P9 and subsequent peaks.
- Lesions of the brainstem and cerebrum delay or abolish the scalp SEP without interfering with clavicular and cervical SEP. The delay causes an increase of the cervical scalp (N13–N20) and clavicular-scalp (N9–N20) conduction times in separate recordings from these points, and an increased separation between N20 and the preceding peaks in recordings between scalp and noncephalic electrodes.

Calculation of peripheral nerve conduction velocity has the advantage of giving a quantitative measure of abnormality that relates directly to pathology. Measurements of central conduction time largely eliminate the effects of peripheral lesions and of changes of temperature, reduce the variation due to different arm length, and help to localize lesions to the peripheral and central segments of the somatosensory pathway. These measurements are also more sensitive to pathology than absolute latencies.

# 17

# Abnormal SEPs to Arm Stimulation

## 17.1 CRITERIA FOR DISTINGUISHING ABNORMAL SEPS

### 17.1.1 Absence of Clavicular, Cervical, and Scalp SEPs

The absence of the N9 peak in recordings from Erb's point, the absence of the N13 peak in recordings from the neck, and the absence of the N20 peak in scalp recordings is abnormal if technical problems are excluded. The absence of the corresponding peaks in recordings between scalp and noncephalic electrodes is likely to be abnormal. The absence of the N13 with preservation of an N20 of normal latency is of doubtful diagnostic significance. Absence of a P13/14 is of importance only in laboratories using methods that consistently produce this peak. An absent N9 with normal N13 and N20 may occur in normal subjects.

### 17.1.2 Slow Peripheral Conduction Velocity

Peripheral conduction velocity is calculated from recordings made with a clavicular elec-

trode by measuring the conduction distance as a straight line between stimulating cathode and the recording electrode at Erb's point and by dividing this distance by the latency from the onset of the stimulus pulse to the peak of the clavicular SEP. Measurement of the peak is preferred over measurement to the onset because peaks are usually better defined than onsets. Peripheral conduction velocity derived with this method does not reflect the maximum velocity but comes closer to the velocity of the majority of the fast-fiber group causing the deflection.

The finding of a decreased peripheral conduction velocity is abnormal unless unexplained by low arm temperature or inaccurate measurement of arm length. The abnormality can be further investigated by recordings of sensory nerve action potentials and of motor nerve conduction velocity. Although these tests generally do not reflect the function of the proximal part of the peripheral nervous system, they can help to identify lesions of the distal part and to distinguish motor from sensory defects.

### 17.1.3 Prolonged Central Conduction Times

Central conduction times are derived by calculating latency differences between peaks recorded at Erb's point, neck, and scalp. Most important are the $N\overline{9}$–$N\overline{13}$ and the $N\overline{13}$–$N\overline{20}$ conduction times. The $N\overline{9}$–$N\overline{20}$ conduction time is usually increased when one of the other two conduction times is increased. In SEPs recorded between scalp and noncephalic reference electrodes, central conduction times are measured as the separation between successive peaks.

The difference between latencies or central conduction times of SEP peaks to stimulation of either side of the body may be abnormal in some patients even when the latencies and central conduction times themselves are not abnormally increased. Abnormally large latency differences suggest a lesion on the side of the pathway responsible for the longer latency. Evaluation of latency asymmetries between SEPs to stimulation of the two sides may increase the sensitivity of the test.

### 17.1.4 Abnormal Latency Differences to Stimulation of Either Side of the Body

The difference between latencies or central conduction times of SEP peaks to stimulation of either side of the body may be abnormal in some patients even when the latencies and central conduction times themselves are not abnormally increased. Abnormally large latency differences suggest a lesion on the side of the pathway responsible for the longer latency. Evaluation of latency asymmetries between SEPs to stimulation of the two sides may increase the sensitivity of the test.

### 17.1.5 Decreased Amplitude

Because SEP amplitude varies considerably and has a non-Gaussian distribution in normal subjects, its use as an indicator of abnormality is limited. Amplitude differences between two SEPs recorded at the same level in response to stimulation of each side may suggest a lesion in the pathway yielding the SEP of lower amplitude. Because lesions in the peripheral and central parts of the sensory pathway selectively reduce peripheral or central SEPs, measurements of amplitude ratios between central and peripheral SEPs may have diagnostic value.

## 17.2 GENERAL CLINICAL INTERPRETATION OF ABNORMAL SEPS TO ARM STIMULATION

Clinical interpretation of arm SEP abnormalities are presented in Table 17.1. Most common arm SEP findings in various disorders are presented in Table 17.2.

### 17.2.1 Technical Problems

SEPs to arm stimulation may be absent at all recording levels due to lack of an effective stimulus, to failure of proper synchronization between stimulator and averager, or to faulty recording electrodes, amplifiers, or averaging channels. The latency of SEPs may be increased at all recording levels due to low body temperature, especially of the stimulated extremity, or due to inaccurate measurement of the length of the extremity.

### 17.2.2 Peripheral and Central Lesions of the Somatosensory Pathway

Absence of the clavicular, cervical, and scalp SEPs indicates a lesion in the peripheral nerve or brachial plexus unless explained by technical problems. Slowing of peripheral conduction velocity, as indicated by an abnormal delay of Erb's point potential, has the same implications.

Absence of the cervical SEP, almost always associated with absence of the scalp SEP, and preservation of the clavicular SEP suggest a lesion involving the spinal cord or roots if technical problems with the cervical and scalp recordings are excluded. An increase of clavicular-cervical ($N\overline{9}$–$N\overline{13}$) and clavicular-scalp ($N\overline{9}$–$N\overline{20}$) conduction times combined with normal peripheral nerve conduction velocity is a rather reliable indicator of a lesion of spinal roots or spinal cord below the lower medulla. An absent scalp SEP with preserved clavicular and cervical SEPs suggests a lesion above the lower medulla if technical problems with the scalp recording

TABLE 17.1. Clinical interpretation of abnormal SEPs to arm stimulation

| Abnormality | Interpretation |
|---|---|
| Absent SEP at all levels | Peripheral nerve lesion; rule out technical problem |
| Absent $\overline{N9}$ with normal $\overline{N13}$ and $\overline{N20}$ | Normal |
| Increased $\overline{N9}$ latency with equally increased $\overline{N13}$ and $\overline{N20}$; decreased peripheral NCV with normal CCT | Peripheral nerve lesion; rule out technical problem |
| Increased $\overline{N9}$–$\overline{N13}$ CT time with normal $\overline{N13}$ amplitude and shape, normal peripheral NCV, normal $\overline{N13}$–$\overline{N20}$ CT | Defect above the brachial plexus and below the lower medulla |
| Absent $\overline{N13}$ and absent or delayed $\overline{N20}$ | Defect above the brachial plexus and below the somatosensory cortex |
| Increased $\overline{N13}$–$\overline{N20}$ CCT with normal $\overline{N9}$–$\overline{N13}$ CT and normal peripheral NCV | Defect above the lower medulla at or below the somatosensory cortex |
| Absent $\overline{N20}$ and a normal $\overline{N9}$–$\overline{N13}$ CT and normal peripheral NCV | Defect above the lower medulla and at or below the somatosensory cortex |
| Decreased peripheral NCV and increased CCT | Combination of peripheral nerve or plexus lesion and central defect |
| Increased latency of all SEPs at all levels | Hypothermia; inaccurate measurement of the distance between stimulating and recording electrodes |

CT = conduction time
NCV = nerve conduction velocity
CCT = central conduction time

are excluded. An increase of clavicular-scalp ($\overline{N9}$–$\overline{N20}$) and cervical-scalp ($\overline{N13}$–$\overline{N20}$) conduction time with normal clavicular-cervical ($\overline{N9}$–$\overline{N13}$) conduction time and normal peripheral conduction velocity indicates a lesion above the lower medulla and below the somatosensory cortex. A combination of decreased peripheral conduction velocity and of increased central conduction times suggests lesions in both the peripheral and the central parts of the somatosensory pathway.

## 17.3 PERIPHERAL LESIONS THAT CAUSE ABNORMAL SEPS TO ARM STIMULATION

In general, peripheral sensory nerve lesions cause abnormalities of SEPs recorded at all levels of the afferent pathway. While complete lesions of peripheral nerves abolish both peripheral and central SEPs, partial peripheral lesions

may make recordings such as sensory nerve action potentials more difficult to obtain than central SEPs. In such instances, scalp SEPs have been used to evaluate nerve lesions.

### 17.3.1 Polyneuropathies

Evaluation of mixed and sensory neuropathies with averaging techniques shows that these neuropathies slow, reduce, or abolish SEPs depending on the degree of neuronal damage rather than on its cause; cervical-scalp ($\overline{N13}$–$\overline{N20}$) conduction time remains intact. Clavicular-cervical ($\overline{N9}$–$\overline{N13}$) conduction time may be increased if the dorsal roots are also involved.

### 17.3.2 Chronic Renal Failure

Scalp SEPs, like VEPs, show an increase of latency and amplitude of long-latency peaks that has no consistent relationship to blood chemistries but tends to return to normal after successful kidney

TABLE 17.2. SEPs to arm stimulation in various disorders

| Disorder | Finding |
|---|---|
| Amyotrophic lateral sclerosis | Normal |
| Anterior spinal artery syndrome | Normal; posterior columns are spared |
| Brachial plexus lesions | Delayed $\overline{N13}$; $\overline{N13}$–$\overline{N20}$ normal |
| Brain death | Absent $\overline{N20}$ and later waves |
| Brainstem stroke | Normal in lateral medullary syndrome. If infarction involves medial lemnisci, absent or delayed $\overline{N20}$ |
| Cervical cord lesion | Loss of $\overline{N13}$, $\overline{N9}$ normal, increased $\overline{N13}$–$\overline{N20}$ |
| Cervical radiculopathy | Usually normal |
| Chronic renal failure | Delay of all peaks and reduced amplitude; $\overline{N13}$–$\overline{N20}$ usually normal |
| Congenital insensitivity to pain | Normal |
| Friedreich's ataxia | Slowed peripheral ($\overline{N9}$–$\overline{N13}$) and often central ($\overline{N13}$–$\overline{N20}$) conduction |
| Guillain-Barré syndrome | Proximal conduction ($\overline{N9}$–$\overline{N13}$) slowed more than distal; normal $\overline{N13}$–$\overline{N20}$ |
| Head injury | Normal or abnormal. A normal response is of moderate prognostic value. |
| Hemispherectomy | Abolish $\overline{N20}$ and later peaks |
| Hepatic encephalopathy | Increased $\overline{N13}$–$\overline{N20}$ |
| Hereditary pressure-sensitive neuropathy | Slowed peripheral conduction |
| Hyperthyroidism | Increased amplitude of $\overline{N20}$ |
| Jakob-Creutzfeldt disease | Variable results; increased or decreased amplitude; may be normal |
| Leukodystrophies | Reduce or abolish $\overline{N13}$; delay or abolish $\overline{N20}$ |
| Minamata disease | Absent $\overline{N20}$; $\overline{N13}$ normal |
| Multiple sclerosis | Increased $\overline{N9}$–$\overline{N13}$ and/or $\overline{N13}$–$\overline{N20}$ |
| Myoclonus epilepsy | Increased $\overline{N20}$ amplitude |
| Parietal lesion | Absent or low amplitude $\overline{N20}$; less delay with cortical than subcortical lesions |
| Perinatal asphyxia | Absence, low amplitude, and/or increased latency of $\overline{N20}$; degree of abnormality correlates with extent of damage |
| Persistent vegetative state | Usually absent or delayed $\overline{N20}$ |
| Polyneuropathy | Delayed peaks; normal $\overline{N13}$–$\overline{N20}$; $\overline{N9}$–$\overline{N13}$ may be prolonged |
| Reye's syndrome | Abnormal early; return of early peaks of good prognostic value |
| Subarachnoid hemorrhage | Low amplitude or absent $\overline{N20}$ |
| Thalamic lesion | Absent or delayed $\overline{N20}$ |
| Thoracic outlet syndrome | Delayed $\overline{N13}$; $\overline{N9}$ may be delayed or low amplitude |
| Tourette's syndrome | Normal |

transplantation. Increased latencies and absence of peaks at different recording points can be explained by peripheral neuropathy in most cases, but involvement of the central somatosensory pathway can not be excluded.

### 17.3.3 Guillain-Barré Syndrome

Clavicular and cervical SEPs are delayed. Sensory conduction is slowed proximally more than distally, cervical radiculopathy may be indicated by an abnormal delay between clavicular and cervical SEP.

### 17.3.4 Charcot-Marie-Tooth Disease, Friedreich's Ataxia, Adie's Syndrome, and Tabes Dorsalis

Peripheral nerve lesions account for most or all SEP abnormalities seen in Charcot-Marie-Tooth disease and for the SEP abnormalities in Adie's syndrome and tabes dorsalis. Friedreich's ataxia may reduce or abolish clavicular SEPs and sensory nerve action potentials (Figure 17.1); cervical and scalp SEPs may be preserved, but show some evidence for a central conduction defect in addition to the peripheral one.

### 17.3.5 Thoracic Outlet Syndrome

Cervical SEPs to ulnar nerve stimulation have been found to be abnormal in patients with cervical ribs and neurological findings. Clavicular SEPs were also abnormal in some of these patients.

### 17.3.6 Brachial Plexus Lesions

Damage to the brachial plexus abolishes or delays SEPs at clavicular and more proximal recording points. Partial plexus lesions, such as lesions of the lower plexus, leave SEPs to stimulation of the unaffected portions intact. SEPs therefore may be useful in the evaluation and management of brachial plexus lesions (Figure 17.2). Complete investigation requires stimulation of several of the nerves contributing to the plexus. Brachial plexus lesions are often associated with cervical root lesions.

### 17.3.7 Carpal Tunnel Syndrome

Scalp SEPs to median nerve stimulation may indicate slowing of conduction even in cases where stimulation of sensory nerve fibers in

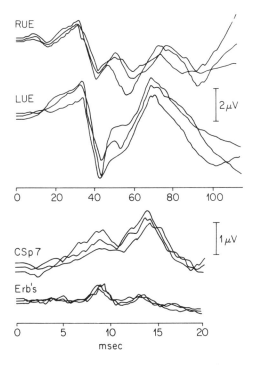

FIGURE 17.1. *SEPs in a patient with Friedreich's ataxia. Top two tracings show delays of the $N\overline{20}$ in the parietal scalp SEPs to stimulation of the contralateral median nerves in the right upper extremity (RUE) and left upper extremity (LUE). Bottom tracings show a cervical SEP (third tracing) with a marginal delay of the $N\overline{13}$ and a clavicular SEP (fourth tracing) with an $N\overline{9}$ of low amplitude, obtained after stimulation of the right arm. Stimulation of the median nerve at the wrist. Recordings from electrodes on the left and right parietal scalp (top two tracings), over the seventh cervical vertebra (third tracing) and Erb's point (bottom tracing). Negativity at these electrodes is plotted upward. Frontal reference electrode. From Nuwer et al. (Ann Neurol 13:20, 1983) with permission by the authors and Little, Brown, and Company.*

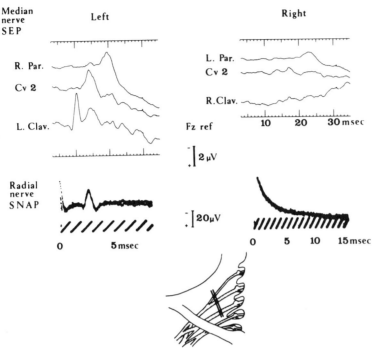

FIGURE 17.2.   *SEPs in a patient with a right brachial plexus traction lesion. Top three tracings on left: Stimulation of the left median nerve and simultaneous recording from right parietal, midline cervical, and left clavicular electrodes with reference to a midfrontal electrode produces normal SEPs. Bottom tracing on left: Stimulation of the left radial nerve produces a normal sensory nerve action potential. Top three tracings on right: Stimulation of the right median nerve produces no peaks in the left parietal, midcervical, and right clavicular recordings. Bottom tracing on right: Stimulation of the right median nerve produces no sensory nerve action potential, probably because of retrograde degeneration. Diagram at bottom: Surgical exploration showed postganglionic lesions of the C5–C7 roots on the right. Positivity at the reference electrode is plotted upward. From Jones et al. (Injury 12:376, 1981) with permission of the authors and John Wright and Sons.*

the digital nerves and recordings from the median nerve showed no sensory nerve action potentials.

## 17.3.8 Congenital Insensitivity to Pain

Cortical SEPs to the usual sensory nerve stimuli remain intact. SEPs to ordinary painful electric tooth pulp stimulation have been found to be altered in one patient, but not in another.

## 17.3.9 Hereditary Pressure-Sensitive Neuropathy

Peripheral conduction defects increase the absolute latencies but not the central conduction times.

## 17.4 LESIONS OF CERVICAL ROOTS AND OF THE CERVICAL CORD THAT CAUSE ABNORMAL SEPS TO ARM STIMULATION

### 17.4.1 Cervical Root Lesions

Lesions of cervical roots are distinguished by preservation of the clavicular SEP and of sensory nerve action potentials, unless there is additional damage to the brachial plexus (Figure 17.3). Postganglionic, but not preganglionic, root damage is followed by retrograde degeneration of sensory nerve fibers and eventual disappearance of the clavicular SEP and of sensory nerve action potentials. Cervical and scalp SEPs are absent in complete avulsions and

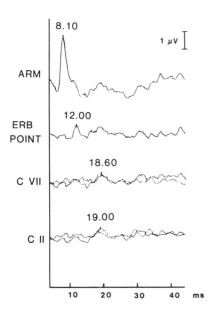

FIGURE 17.3.   *SEPs in a patient with C5–C6 root lesions. The nerve action potential (ARM) and clavicular SEP (ERB POINT) have normal latency by the standards of the particular laboratory. Lower (CVII) and upper (CII) cervical SEPs are delayed. The clavicular and cervical SEPs are also attenuated. Stimulation of the median nerve at the wrist on the involved side. Recording from electrodes on the arm, at Erb's point, and over C7 and C2. Negativity at these electrodes is plotted upward. Midfrontal reference electrode. From Synek and Cowan (Neurology 32:1347, 1982) with permission of the authors and Modern Medicine Publications, Inc.*

may be delayed or reduced in incomplete lesions (Figure 17.3) such as spondylotic radiculopathy. Herpes zoster radiculitis delays the cervical SEP and may reduce the amplitude of the clavicular SEP. As with plexus lesions, clear definition of the level of the lesion requires multiple nerve stimulation.

## 17.4.2 Cervical Cord Lesions

Although lower spinal cord lesions must be studied with SEPs to leg stimulation, lesions of the cervical cord may be detected with SEPs to arm stimulation. These lesions may cause abnormal cervical and scalp SEPs with pre-

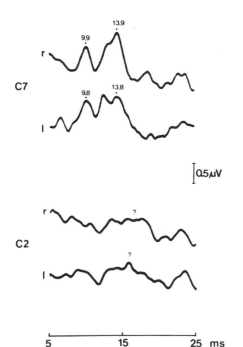

FIGURE 17.4.   *Abnormal SEPs in a patient with cervical myelopathy at C4/C5. Stimulation of the right (r) and left (l) median nerve produced normal SEPs at C7 (top two tracings) but no definite SEPs at C2 (bottom two tracings). Negativity at the neck electrodes is plotted upward. Midfrontal reference electrode. From Stöhr et al. (EEG Clin Neurophysiol 54:257, 1982) with permission of the authors and Elsevier Scientific Publishers Ireland Ltd.*

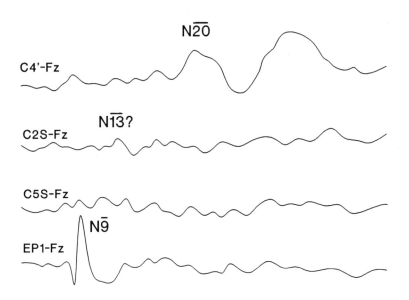

FIGURE 17.5. *Abnormal SEPs to stimulation of the median nerve in a patient with MS. Stimulation of the left median nerve produces a clavicular SEP of normal latency (bottom tracing). Recordings at C5 and C2 (middle two tracings) show no definite peaks in the normal latency range. The scalp recording shows an abnormally delayed N20 indicating a defect above the level of the brachial plexus. This 31-year-old man has MS without clinically detectable sensory involvement.*

served clavicular SEPs. Such abnormalities have been found in many cases of cervical spondylotic myelopathy with or without radiculopathy (Figure 17.4) and in some cases of subacute combined degeneration of the spinal cord due to vitamin $B_{12}$ deficiency, cervical cord injury, tumors, syringomyelia, and hydromyelia. In contrast, infarcts in the distribution of the anterior spinal artery, which do not involve the posterior columns, leave the SEP intact.

## 17.5 LESIONS OF THE BRAINSTEM AND CEREBRAL HEMISPHERES THAT CAUSE ABNORMAL SEPS TO ARM STIMULATION

### 17.5.1 Multiple Sclerosis

Because MS may interrupt fibers at any level of the central somatosensory pathway, it may produce abnormalities of the cervical and scalp SEPs (Figure 17.5) or of the scalp SEP only (Figure 17.6); sometimes only the cervical SEP is abnormal. Delays of these SEPs lead to prolongation of clavicular-cervical (N9–N13), clavicular-scalp (N9–N20), and cervical-scalp (N13–N20) conduction times. The chance of finding SEP abnormalities in MS is increased by using SEPs to leg stimulation which test the entire length of the spinal cord and are more often affected by the diffuse and scattered lesions of MS.

SEPs to arm stimulation are abnormal in the majority of patients with MS, more often in definite than in probable or possible cases. Although the incidence of abnormal SEPs is very high in patients who show defects of vibration, position, and touch sensation in the stimulated limb, a substantial portion of MS patients with abnormal SEPs have normal sensation and may have clinically silent plaques. Abnormal SEPs can be found in patients presenting with only optic neuritis. The incidence of SEP abnormalities in patients with MS increases with progression of the disease, but in patients without relapses, the SEP was found to remain stable. Steady-state scalp SEPs may be abnormal in MS. A normal SEP to median nerve stimulation distinguishes acute transverse myelitis of other causes, involving the spinal cord below the cervical level, from the myelopathic form of MS, which may be associated with abnormal SEPs to arm stimulation due to clinically silent lesions at or above the cervical cord.

The usefulness of the SEP in the diagnosis of MS has been evaluated by comparing the proportions of abnormal SEPs with those of abnormal VEPs and AEPs in patients with pos-

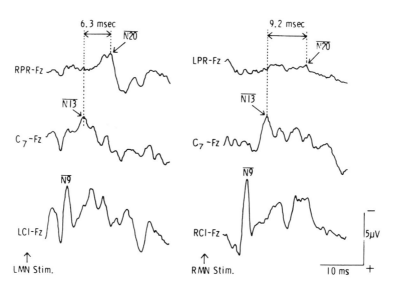

FIGURE 17.6. *SEPs to median nerve stimulation in a patient with MS. Stimulation of the left median nerve (left half of figure) produces normal clavicular (LC1-Fz), cervical (C₇-Fz), and right parietal scalp (RPR-Fz) SEPs (bottom to top). Stimulation of the right median nerve (right half of figure produces normal clavicular and cervical SEPs, but the scalp SEP is attenuated indicating a conduction defect above the level of the spinal cord. This 21-year-old woman suffers from MS producing complete proprioceptive sensory loss of the right hand. From Shibasaki et al. (J Neurol Sci 57:441, 1982) with permission of the authors and Elsevier Biomedical Press B.V.*

sible, probable, and definite MS. In general, SEPs and VEPs are more effective than BAEPs. SEPs are abnormal more often than VEPs if both arm and leg stimulation are used. Studies using SEPs to arm stimulation alone have reported the SEP to be abnormal more frequently than the VEP in patients with possible or probable MS or with definite and possible MS. However, some studies of the SEP to arm stimulation found the VEP to be more effective in most diagnostic groups.

## 17.5.2 Brainstem Strokes and Tumors

Strokes and tumors of the brainstem cause scalp SEP abnormalities in many cases. In the "locked-in syndrome" the SEP may be abnormal if the pontine lesion extends into the medial lemnisci of the tegmentum. In contrast, infarcts causing the lateral medullary syndrome of Wallenberg or the peduncular syndrome of Weber, neither of which usually involves lemniscal fibers, leave the SEP intact.

## 17.5.3 Thalamic Lesions

Lesions of the thalamus may cause delays or reductions of the N20–P30 peaks of the scalp SEP; the N13, and often also the P13/14, remains intact (Figure 17.7). Abnormal scalp SEPs were found to result from lesions that involve the nucleus ventralis posterolateralis and encroach on the nucleus ventralis posteromedialis, but not from lesions of the nucleus ventralis lateralis. Scalp SEPs are likely to be abnormal in cases of thalamic lesions that affect position, vibration, and touch sensation, including cases of the thalamic syndrome of Dejerine and Roussy that show impairment of these modalities.

## 17.5.4 Parietal Infarcts and Tumors

Mass lesions involving the primary sensory receiving areas produce abnormalities of the scalp SEP which are usually associated with contralateral sensory defects; only a few patients with sensory defects have normal SEPs. Most

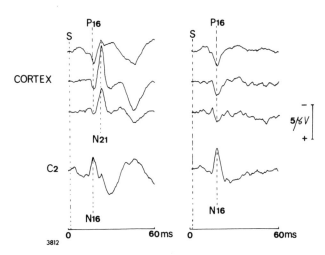

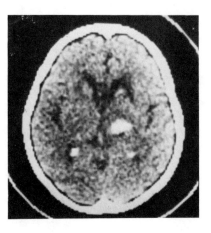

FIGURE 17.7.   *SEPs in a patient with a right thalamic hemorrhage. Stimulation of the right median nerve (left half of figure) produces normal scalp (top three tracings) and cervical (bottom tracing) SEPs. Stimulation of the left median nerve (right half of figure) produces only an early positive scalp peak corresponding with the negative neck SEP. The peaks N16, P16, and P21 in this figure correspond to the normal N13, P13/14, and N20. Recordings from electrodes 2 cm behind Cz (top tracing), 5 cm (second tracing), and 7 cm (third tracing) lateral to that electrode, and from an electrode on the C2 spinous process (bottom tracing). Negativity at these electrodes is plotted upward. Reference electrodes on interconnected ears for top three tracings and at Fz for bottom tracing. This 70-year-old woman has a complete left hemiplagia and hemianesthesia. The CT scan shows a hemorrhage in the right internal capsule and thalamus. From Mauguière and Courjon (Ann Neurol 9:607, 1981) with permission of the authors and Little, Brown, and Company.*

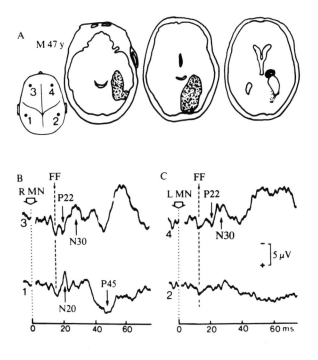

FIGURE 17.8.   *Scalp SEPs in a patient with a right parietal lesion. Stimulation of the right median nerve (R MN) in (B) produces normal prerolandic (3) and parietal (1) SEPs, including an early far-field peak (FF). Stimulation of the left median nerve (L MN) in (C) produces a fairly normal prerolandic SEP (4) but no clear parietal peaks after the far-field peak (2). Recording with scalp electrodes as indicated in the diagram (A). Negativity at the scalp electrodes is plotted upward. Ipsilateral ear reference electrodes. This 47-year-old man had suffered a sudden cerebrovascular lesion five years earlier and has residual left astereognosis, loss of graphesthesia, position sense and two-point discrimination, and left upper quadrantanopsia. The CT scan showed an area of reduced density in the right parieto-occipital and posterior thalamic regions (A). From Mauguière et al. (Brain 106:271, 1983) with permission of the authors and Oxford University Press.*

often, parietal lesions reduce or abolish the N20–P30 and later peaks while leaving earlier peaks intact (Figure 17.8). An increase of latency is less characteristic of cerebral lesions than of afferent pathways. Central conduction time has been shown to increase transiently with pathological reductions in cerebral blood flow. Amplitude reductions of the N20 peak in patients paralyzed by strokes have been said to suggest an unfavorable prognosis.

### 17.5.5 Lesions in Other Cerebral Areas

Cerebral lesions outside the primary sensory area may reduce parietal SEP peaks of latencies longer than those of N20 and P30. Lesions

in the frontal area may selectively reduce SEP peaks recorded from that area (Figure 17.9). Lesions outside the primary sensory area may increase amplitude and duration of the parietal SEP, whereas chronic parietal lesions may increase the SEP in the frontal area. Unilateral cerebral lesions alter SEPs recorded with different latency on both sides of the head after stimulation of the contralateral arm, but these lesions do not affect the SEPs produced on both sides by stimulation of the ipsilateral arm.

### 17.5.6 Hemispherectomy

Removal of one hemisphere eliminates scalp SEP peaks of medium latency recorded over

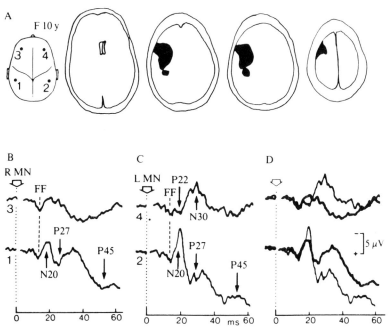

FIGURE 17.9. *Scalp SEPs in a patient with a left prerolandic lesion. Stimulation of the right median nerve (R MN) in* (B) *produces a normal parietal SEP (1), including an early far-field potential (FF); the prerolandic recording shows no peaks after the far-field peak (3). Stimulation of the left median nerve (L MN) in* (C) *produces normal parietal (2) and prerolandic (4) SEPs. The tracings of* (B) *and* (C) *are superimposed in* (D)*, the heavier lines representing the involved side. Recordings from scalp electrodes as indicated in diagram* (A)*. Negativity at the scalp electrodes is plotted upward. Ipsilateral ear reference electrodes. This 10-year-old girl suffered an intracerebral hemorrhage from an arteriovenous malformation one year earlier and has residual right spastic hemiplegia and moderate dysphasia and dysarthria but no sensory abnormalities. The CT scan shows a left rolandic lesion* (A)*. From Mauguière et al. (Brain 106:271, 1983) with permission of the authors and Oxford University Press.*

both sides of the head in response to stimulation of the arm contralateral to the hemispherectomy. Early peaks, generated by preserved subcortical structures, and late peaks, presumably generated in the ipsilateral hemisphere via extralemniscal afferents, may be preserved.

## 17.5.7 Diffuse Cerebral Disorders with Reduced Consciousness

### 17.5.7.1 Cerebral death

The precise role of SEPs, like that of other EPs in evaluating cerebral death, has not been clarified. In general, patients who fulfill the criteria for cerebral death and have no electrocerebral activity on EEG show no $\overline{N20}$ and $\overline{P30}$ on scalp SEP, even though subcortically originating peaks may persist in the scalp and neck SEPs (Figure 17.10). The persistence of cervical SEPs in cases without cortical SEPs proves that the sensory input has reached the central nervous system so the absence of the higher SEPs can safely be attributed to absent cerebral functioning. Patients showing alpha frequency coma pattern on EEG may retain SEPs until the EEG further deteriorates or disappears. However, the absence of central SEPs does not correspond precisely with brain death: An $\overline{N20}$ has been reported to persist in the SEP of a case of cerebral death diagnosed with angiography but without EEG recording.

### 17.5.7.2 Head injury

SEPs alone or in combination with other EPs have been used to evaluate the severity of cerebral injuries and to make an early prognosis. Serial recordings showing a reduction in central conduction time or an increase in the number of peaks of the scalp SEP generally indicated an imminent return of function.

### 17.5.7.3 Chronic vegetative state

Various combinations of scalp and cervical SEP abnormalities have been described. In general, the cervical SEP is preserved; complete absence of cortical SEPs suggests a poor outcome.

### 17.5.7.4 Reye's syndrome

SEPs may be of prognostic value. Scalp potentials are nearly or completely abolished initially. Early recovery of short-latency peaks indicates survival; progressive recovery of peaks with latencies over 100 msec precedes satisfactory clinical recovery.

### 17.5.7.5 Subarachnoid hemorrhage

Central conduction time has been found to be increased in patients with aneurysmal subarachnoid hemorrhage in poor condition and with poor prognosis.

### 17.5.7.6 Perinatal asphyxia

Scalp SEPs have been reported to be abnormal in the majority of asphyxiated newborn infants, showing absence, reduced number, or low amplitude of peaks and increased latency. The degree of these abnormalities corresponded with the severity of the asphyxia. Patients with normal SEPs can expect good neurologic outcomes, but the majority of patients with abnormal or absent SEPs have poor neurologic outcome.[26,38]

### 17.5.7.7 Surgical monitoring

Scalp SEPs to arm stimulation have been used to monitor the condition of the brain during carotid endarterectomy and aneurysmal surgery.[14,33] Loss of median SEP intraoperatively is highly predictive of subsequent neurologic deficit.[42] Preservation of normal SEPs was only rarely followed by development of neurologic deficit. Pure motor hemiparesis was associated with normal SEPs in one patient undergoing surgery for intracranial aneurysm.[51]

Exploration of the brachial plexus is also sometimes monitored by arm SEPs.

Many inhalation anesthetics, including halothane, enflurane, and isoflurane produce a reduction in the amplitude of the cortical SEPs. Nitrous oxide, in contrast, has minimal effect on the potentials. Subcortical potentials are affected to a much lesser extent by inhalation anesthetics.

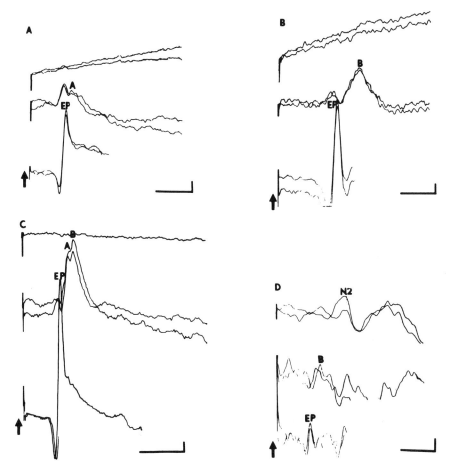

FIGURE 17.10. *SEPs from three brain-dead patients (A–C) and a comatose but not brain-dead patient (D). The three tracings for each patient represent scalp (top), neck (middle), and clavicular (bottom) recordings. The three brain-dead patients have no scalp SEPs. Neck and clavicular SEPs are preserved although the neck SEP in patient (A) is abnormal. The comatose patient in (D) has cervical and scalp peaks preserved although distorted and delayed, possibly due to hypothermia of less than 92°F. Stimulation of the median nerve at the wrist. EP = Erb's point potential; A, B = cervical peaks; N2 = N$\overline{20}$. Negativity at the scalp, neck, and clavicular electrodes is plotted upward. Midfrontal reference electrode. Calibration bars are 10 msec and 0.25 μsecV. From Goldie et al. (Neurology 31:248, 1981) with permission of the authors and Modern Medicine Publications, Inc.*

## 17.5.8 Diffuse Cerebral Disorders without Reduced Consciousness

### 17.5.8.1 Juvenile myoclonic epilepsy, Jakob-Creutzfeldt disease, Ramsay-Hunt syndrome, and other conditions associated with myoclonus

The scalp SEP is greatly increased in amplitude in many patients with juvenile myoclonic epilepsy (JME), especially if myoclonus is present at the time of the recording, and is similar to the EEG spike that precedes spontaneous myoclonic jerks in this disorder.[71,73]

In Jakob-Creutzfeldt disease, the amplitude of the scalp SEP has been found to be increased in some studies and increased or decreased in others.[46,70,77] No peaks after N$\overline{20}$ were found in one study. The cervical SEP in another case report was normal.

The SEP amplitude has been reported to be increased in the dyssynergia myoclonica of Ramsay-Hunt, in familial startle disease, in a few cases of postanoxic myoclonus, and in one case of unilateral myoclonus after stroke.

### 17.5.8.2 Friedreich's and other hereditary cerebellar ataxias, familial spastic paraplegia, and amyotrophic lateral sclerosis

Friedreich's ataxia often causes a decrease of amplitude and temporal dispersion of the N20–P30 of the scalp SEP, suggesting cerebral conduction defects in addition to the peripheral conduction defects. Abnormal SEPs were found in olivopontocerebellar degeneration and in some cases of hereditary cerebellar ataxia and familial spastic paraplegia, but not in isolated cases of cerebellar ataxia. A normal SEP generally distinguished patients with amyotrophic lateral sclerosis although an abnormal SEP has been reported in one case.

### 17.5.8.3 Huntington's chorea

The scalp SEP has low amplitude and may have slightly increased latency in patients and many subjects at risk.

### 17.5.8.4 Wilson's disease

The SEP may show reduction of scalp SEP peaks following the N20 or increased central conduction time.

### 17.5.8.5 Myotonic dystrophy

A few patients show increased central conduction times.

### 17.5.8.6 Leukodystrophies

Pelizaeus-Merzbacher disease, adrenoleukodystrophy, and metachromatic leukodystrophy may reduce or abolish the cervical SEP and delay or abolish the scalp SEP.

### 17.5.8.7 Down's syndrome

The amplitude and latency of scalp SEPs may be increased.

### 17.5.8.8 Hepatic encephalopathy

Central conduction times may be increased.

### 17.5.8.9 Hyperthyroidism

The scalp SEP, like the VEP, has been reported to have increased amplitude but essentially normal latency.

### 17.5.8.10 Minamata disease

The N20 has been reported to be absent; clavicular and cervical SEPs were preserved.

### 17.5.8.11 Tourette's syndrome

SEPs have been found to be normal.

### 17.5.8.12 Drug effects

Like other EPs, SEPs show diverse changes of late peaks of the scalp in patients taking drugs that influence behavior, whereas the early peaks commonly used in clinical diagnosis remain unaffected. Depressant drugs generally increase latency and decrease amplitude.

### 17.5.8.13 Psychiatric disorders

SEPs, like other EPs, have been reported to differ from normal in patients with schizophrenia and affective psychoses, but no specific diagnostic SEP abnormalities have been isolated. Schizophrenia tends to reduce the amplitude of scalp SEP peaks of over 100 msec latency; in chronic paranoid patients, a negative peak at 60 msec was found to be increased. SEP measures may distinguish chronic schizophrenia from psychiatric depression.

Hysterical absence of pain sensation has been reported not to cause SEP abnormalities. Hypnotically induced anesthesia produced no significant changes in the scalp SEP in one study but reduced SEP amplitude in another.

### 17.5.8.14 Parkinsonism

Patients with progressive supranuclear palsy (PSP) may have slowed conduction through the brainstem.[64] Patients with idiopathic Parkinson's disease have typically had normal SEPs.[19,60]

# 18

# Normal SEPs to Leg Stimulation

## 18.1 NORMAL SEPS AT DIFFERENT RECORDING SITES

SEPs to leg stimulation may be produced by stimulating the posterior tibial nerve at the ankle (Figure 18.1) or the common peroneal nerve at the knee (Figure 18.2) and by recording from the lumbothoracic spine and the scalp. Recordings from the neck do not reliably show SEPs in response to leg stimulation. Nerve action potentials can be recorded at the knee when the posterior tibial nerve is stimulated. Other nerves are used occasionally for stimulation. Like SEPs to arm stimulation, SEPs to leg stimulation vary depending on subject variables, stimulus characteristics, and recording parameters (Table 18.1).

### 18.1.1 Popliteal Fossa Potential

To record the action potential of tibial nerve fibers at the knee after stimulation of the posterior tibial nerve at the ankle, a popliteal fossa electrode is placed 4–6 cm above the popliteal crease, midway between the combined tendons of the semimembranosus and semitendinosus muscles medially and the tendon of the biceps femoris laterally. These tendons can be brought out by having the subject bend the knee against resistance. The popliteal electrode may be referred to an electrode on the medial surface of the knee.

Like the clavicular SEP, the popliteal fossa potential consists of a major negative peak that may be preceded and followed by smaller positive peaks (Figure 18.3). The latency, measured to the negative peak, depends on the distance between stimulus and recording electrodes and amounts to about 9–10 msec in normal adults.

### 18.1.2 Lumbar and Low Thoracic SEPs

Recording electrodes may be placed over the spinous processes of L3, T12, and T6 (electrodes L3S, T12S, and T6S). The process of L3 lies above a line connecting both iliac crests. Reference electrodes may be placed 4 cm above each of these electrodes. Alternatively, an electrode over the iliac crest or at Fz may be used as a common reference for the spinal electrodes. All three spinal electrode pairs may be

193

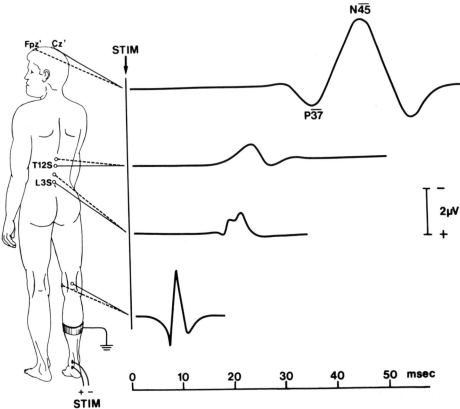

FIGURE 18.1. *Schematic diagram of normal SEPs to stimulation of the posterior tibial nerve at the ankle. Tracings, from bottom to top, show popliteal fossa potential, lumbar and low thoracic spinal potentials, and scalp SEP. Negativity at the electrodes connected with a solid line in the diagram at the left is plotted upward.*

used, in addition to a scalp electrode pair, in four-channel recordings of the SEPs to common peroneal nerve stimulation. Only the lower two pairs are used in four-channel recordings of SEPs to posterior tibial nerve stimulation, the other two channels being allocated to scalp and popliteal fossa recordings (Table 18.1).

Lumbar SEPs have a negative peak that may be preceded by a small positive peak. These peaks may be followed by a second negative peak that is best recorded at a slightly higher level. The latency of the lumbar peaks depends on the stimulation and recording sites, the length and temperature of the leg, and the peripheral conduction velocity. The first negative peak recorded at the lumbosacral area normally has a latency of about 17–21 msec

after stimulation of the posterior tibial nerve at the ankle (Figure 18.3) and of about 9–12 msec after stimulation of the common peroneal nerve at the knee (Figure 18.4). This peak is used to calculate the lumbar-scalp conduction time.

The small positive peak, and perhaps also the first negative wave, of the lumbar SEP probably originates from the dorsal roots of the cauda equina. The second negative peak probably represents postsynaptic cord elements; ventral root discharges can be recorded with special methods. The generators of these SEPs have been investigated further with epidural and intrathecal recordings.

The low thoracic SEP consists of one or more peaks that probably reflect activity of intramedullary continuations of the dorsal root fibers followed by synaptic and postsynaptic spinal activity.

### 18.1.3 The Scalp SEP

Scalp SEPs to leg stimulation are recorded with an electrode 2 cm behind the Cz electrode position of the International 10–20 System (Cz'), and midway between Fpz and Fz positions of the 10–20 System (Fpz'). The derivation used is Cz'–Fpz'.

Stimulation of the posterior tibial nerve at the ankle produces a parietal P$\overline{37}$–N$\overline{45}$ complex (Figure 18.3); stimulation of the common peroneal nerve at the knee and recording from the

scalp produces a positive-negative complex, called P$\overline{27}$–N$\overline{35}$ (Figure 18.4). The actual normal values vary considerably with body height and other factors. In some instances, the first positive peak is preceded by a small negative potential. The early positive-negative complex may be followed by later, more variable peaks.

The distribution of the positive-negative complex is restricted to the postcentral area; later peaks have a wider distribution. Although usually recorded at the midline or slightly contralateral to the stimulated leg, the scalp SEP to leg stimulation has been reported to have higher amplitude over the ipsilateral scalp, suggesting that the electric field generated in the contralateral leg area of the parietal cortex is so oriented

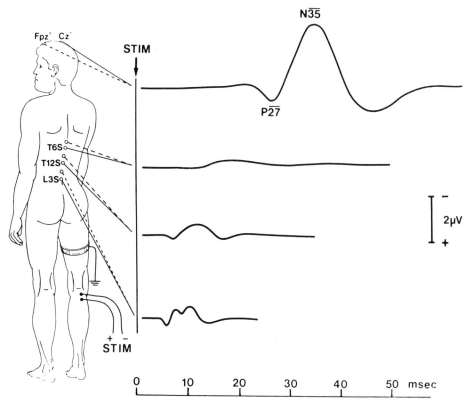

FIGURE 18.2. *Schematic diagram of the normal SEPs to stimulation of the common peroneal nerve at the knee. Tracings, from bottom to top, show lumbar, low thoracic, and middle thoracic spinal potentials, and scalp SEP. Negativity at the electrodes connected with a solid line in the diagram at the left is plotted upward.*

TABLE 18.1.   SEPs to stimulation of the common peroneal nerve (CPN) at the knee or the posterior tibial nerve (PTN) at the ankle

A. Subject variables as in Table 16.1 except:
   1. Age: Conduction velocities vary in children; age ranges for control groups have not yet been standardized.
B. Stimulus as in Table 16.1 except:
   1. Electrode placement:
      a. For PTN stimulation: Cathode behind medial malleolus; anode 3 cm distal
      b. For CPN stimulation: Cathode in lateral popliteal fossa; anode 3 cm distal
   2. Intensity
      a. For PTN stimulation: Above threshold for twitch causing plantar toe flexion
      b. For CPN stimulation: Above threshold for muscle twitch eversion of the foot
   3. Unilateral stimulation: Each side should be stimulated separately in routine recordings. Simultaneous bilateral stimulation is used during surgical monitoring to enhance SEP amplitude.
C. Recording. As in Table 16.1 except:
   1. Electrode placement
      a. For PTN stimulation: Over the tibial nerve in the upper middle popliteal fossa (PF) with a reference electrode on the medial surface of the knee; over spinous processes of L3 and T12 vertebrae and on the scalp as for CPN stimulation except for the top spinal pair; ground electrode on the calf.
      b. For CPN stimulation: Over L3, T12, and T6 spinous processes (L3S, T12S, T6S) with reference electrodes 4 cm rostral to each of these three electrodes; on the scalp 2 cm posterior to Cz' with a reference electrode midway between Fpz and Fz (Fpz'); ground electrode at midthigh level.
   2. Montages
      a. For PTN stimulation:
         Channel 1: Cz'–Fpz'
         Channel 2: L3S–electrode 4 cm rostral
         Channel 3: L3S–electrode 4 cm rostral
         Channel 4: PF–medial surface of knee
      b. For CPN stimulation:
         Channel 1: Cz'–Fpz'
         Channel 2: T6S–electrode 4 cm rostral
         Channel 3: T12S–electrode 4 cm rostral
         Channel 4: L3S–electrode 4 cm rostral
   3. Number of responses averaged: 1,000–4,000
   4. Sweep length
      a. For PTN stimulation: 60–80 msec; 100–200 msec for delayed SEPs
      b. For CPN stimulation: 40–60 msec; 100–200 msec for delayed SEPs
D. Analysis
   1. Normal peaks
      a. For PTN stimulation: PF potential, L3 and T12 spine potentials, $\overline{P37}$ and $\overline{N45}$ peaks of scalp SEPs
      b. For CPN stimulation: L3, T12, and T6 spine potential, $\overline{P27}$ and $\overline{N35}$ peaks of scalp SEPs
   2. Criteria of abnormal
      a. Absence of all spine and scalp SEPs in recordings including 100–200 msec sweeps
      b. Extremely slow peripheral conduction velocity from stimulus cathode to PF peak and to L3 peak for PTN stimulation and from stimulus cathode to L3 peak for CPN stimulation
      c. Abnormally slow central conduction velocity from L3 and T12 to scalp $\overline{P37}$ peak for PTN stimulation and from L3, T12, and T6 spine potential to scalp $\overline{P27}$ peak for CPN stimulation

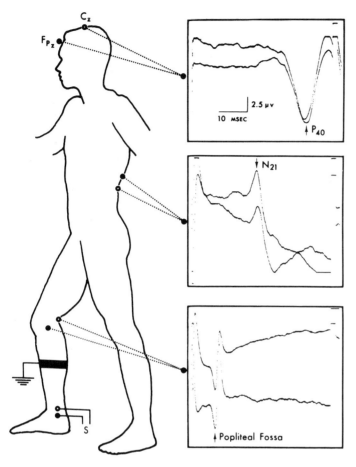

FIGURE 18.3. *Normal SEPs to stimulation of the posterior tibial nerve at the ankle. Recordings from the popliteal fossa (bottom tracings) show a negative peak at about 10 msec preceded by a positive peak. Recordings from the low thoracic spine (middle tracings) contain a negative peak at about 21 msec (N21). Scalp recordings (top tracings) show a P40 peak. Two tracings are superimposed for each recording site. Negativity at the recording electrodes marked with open circles is plotted upward. From Eisen and Odusote (EEG Clin Neurophysiol 48:253, 1980) with permission of the authors and Elsevier Scientific Publishers Ireland Ltd.*

that it produces higher potential gradients over the ipsilateral hemisphere. Recordings between a central scalp electrode and extracranial reference electrodes show small and variable earlier peaks, probably representing far-field SEPs from subcortical structures.

## 18.2 SUBJECT VARIABLES

### 18.2.1 Age

Measurements of SEPs at different levels of the spinal cord in infants and children have suggested that conduction velocities of peripheral nerves and spinal cord in newborns are about half those of adults; adult velocities are reached at the age of about 3 years for peripheral conduction and at about 4 to 5 years for cord conduction. Although peripheral conduction velocity decreases steadily throughout adulthood, spinal cord conduction velocity has been reported to show little change up to the age of 60 years and then to decrease sharply.

### 18.2.2 Leg Length, Body Height, Temperature, and Sensation

These variables have the same effect on leg SEPs as on arm SEPs, except that body height is more important for leg SEPs than for arm SEPs and therefore must be taken into account by calculating central conduction velocities.

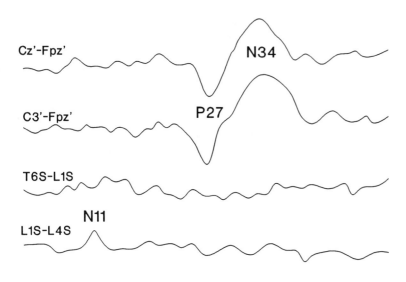

Cz'–Fpz'   N34

C3'–Fpz'   P27

T6S–L1S

L1S–L4S   N11

FIGURE 18.4.   *Normal SEPs to stimulation of the common peroneal nerve at the knee. Recordings from lumbar and thoracolumbar electrodes show small negative peaks at about 11 and possibly 13 msec (bottom two tracings). Scalp recordings show P$\overline{27}$ and N$\overline{34}$ peaks (top two tracings). The beginning of the tracing is delayed by a few milliseconds against the stimulus to eliminate the stimulus artifact. Negativity at the first electrode in the pairs indicated at the left margin is plotted upward.*

## 18.3 EFFECT OF STIMULUS PARAMETERS

### 18.3.1 Stimulus Electrode Placements

Stimulus electrode placements for each nerve are fairly well standardized. For stimulation of the posterior tibial nerve, one electrode is placed midway between the medial border of the Achilles tendon and the posterior border of the medial malleolus. This electrode is connected to the negative pole of the stimulator and serves as the cathode. Another electrode is placed 3 cm distal to the cathode and connected to the positive pole of the stimulator, serving as the anode. A band electrode around the calf or a plate electrode on the calf is used as a ground lead (Figure 18.1). For stimulation of the common peroneal nerve at the knee, the cathode is placed over the lateral part of the popliteal fossa, just medial to the tendon of the biceps femoris and below the leg crease. The tendon of the biceps femoris can be easily seen when the subject bends the knee against a resistance. The anode is placed 3 cm distal to the cathode. A plate or band electrode on the midthigh is used as a ground lead (Figure 18.2). Purely sensory fibers may be excited by stimulating the sural nerve at the lateral malleolus, the superficial peroneal nerve a handbreadth above the lateral malleolus, or the saphenous nerve above and anterior

to the medial malleolus. Segmental input from L2–S1 can be tested by stimulating various nerve branches; L2, L3 can be tested by stimulating the lateral femoral cutaneous nerve at the thigh; L3, by stimulating the saphenous nerve at the medial side of the knee; L4, by stimulating the saphenous nerve below the inner malleolus; L5, by stimulating the superficial peroneal nerve on the lateral aspect of the leg; and S1, by stimulating the sural nerve above the ateral malleolus. Segments not well represented by sensory branches may be studied by direct stimulation of the skin of the corresponding dermatomes.

### 18.3.2 Stimulus Intensity, Rate, and Duration

The motor threshold for posterior tibial nerve stimulation at the ankle is marked by plantar flexion of the toes; the threshold for stimulation of the common peroneal nerve at the knee is indicated by eversion of the foot. Procedures for selecting stimulus intensity for mixed nerves in case of peripheral nerve lesions, and for sensory nerve stimulation, are described in chapter 15.

An increase of stimulus intensity does not change the latency of the lumbar SEP but increases its amplitude to a maximum.

Stimulus rate is 4–7/sec. Stimulus duration is usually 200–300 μsec.

## 18.4 RECORDING PARAMETERS

Most recording parameters, including electrode placements, have been described and summarized (Table 18.1). About 1,000–4,000 responses are averaged for SEPs to stimulation of the posterior tibial and the common peroneal nerve. Larger numbers may be needed for the small SEPs to stimulation of small nerves or dermatomes. The sweep length is 60–80 msec for posterior tibial nerve stimulation and 40–60 msec for common peroneal stimulation. Longer sweeps are needed for abnormally delayed SEPs. With longer sweeps, the memory capacity of older EP machines may be exceeded.

## 18.5 GENERAL STRATEGY OF STIMULATING AND RECORDING

SEPs to posterior tibial nerve stimulation recorded from the popliteal fossa, lumbothoracic spine, and scalp allow distinction of lesions in three segments:

1. lesions of the distal peripheral nerve affect the popliteal fossa potential and later SEPs;
2. lesions of the proximal peripheral nerve and cauda equina leave the popliteal fossa potential intact but affect lumbothoracic and scalp SEPs; and
3. lesions of the spinal cord, brainstem, or cerebral hemispheres leave the lumbar SEP intact but delay or abolish the scalp SEP.

SEPs to stimulation of the common peroneal nerve, recorded at lumbothoracic and scalp electrodes, can distinguish lesions only in segments (2) and (3). In any segment, lesions produce SEP abnormalities only if they involve specific parts of the somatosensory pathway, namely, the sensory fibers contained in peripheral nerves, plexus, or spinal roots, the dorsal columns of the spinal cord, the medial lemniscus of the brainstem and diencephalon, the nucleus ventralis posterolateralis of the thalamus, the thalamic radiations, and the leg area of the somatosensory cortex.

A lesion in any segment either abolishes or delays all potentials generated proximally. The increase in latency is best referred to the length of the segment and expressed in terms of peripheral and central conduction velocity. Whereas peripheral conduction velocity approximately equals the speed of conduction of the majority of the fast fibers in a peripheral nerve, the term *central conduction velocity* is not entirely appropriate in this context because it is used to denote values that represent not only axonal conduction but also synaptic transmission. Nevertheless, it is necessary to use a measure that relates central conduction time to the conduction distance. In contrast to the rather short central conduction distances for SEPs to arm stimulation, the central conduction distances for SEPs to leg stimulation, extending from the lower spine to the vertex, are fairly long and vary considerably with the subject's height; these variations affect both the latencies and central conduction times of leg SEPs.

Recordings from electrodes over the cervical and upper thoracic spine do not usually yield SEPs in adults, although they often show well-defined peaks in infants and then permit calculation of spinal conduction times separately from brainstem and cerebral conduction times. In adults, conduction times through the higher segments of the somatosensory pathway have been estimated indirectly from F and M waves or H responses recorded in conjunction with the scalp SEP. Conduction times between lumbar and thoracic electrodes are too variable in normal patients to be of clinical value for estimating conduction velocity between cord segments. Epidural, intrathecal, and esophageal recordings suggest spinal conduction velocities of 35–85 m/sec.

The level of a lesion may be evaluated by recording both arm and leg SEPs. Lesions of the upper cervical cord or higher structures may cause abnormalities of both kinds of SEPs, whereas lesions below the upper cervical cord may render abnormal the SEPs to leg stimulation while leaving intact the SEPs to arm stimulation. More detailed information on the level of lesions can be obtained with the painstaking study of segmental input from various nerves and dermatomes.

# 19

# Abnormal SEPs to Leg Stimulation

## 19.1 CRITERIA DISTINGUISHING ABNORMAL SEPS

The absence of SEPs at all recording levels, namely, the popliteal fossa, lumbar and thoracic spinal column, and the scalp, is abnormal if technical problems are excluded and if abnormally delayed peaks have been looked for with sweeps of up to 200 msec. However, the absence of spinal SEPs can not be considered abnormal if scalp SEPs of normal latency can be recorded. The absence of scalp SEPs with preserved normal spinal SEPs, although consistent with a lesion above the spinal recording level, can not be considered a definite abnormality because scalp SEPs often have very low amplitude, making them difficult to record.

### 19.1.1 Slow Peripheral Conduction Velocity

Peripheral conduction velocity of the common peroneal nerve is calculated by measuring the straight-line distance between the stimulating cathode in the popliteal fossa and the L3S recording electrode, and dividing this distance by the latency from the leading edge of the stimulus pulse to the peak of the L3 potential.

Peripheral conduction velocities may be calculated for three segments of nerve with posterior tibial nerve stimulation:

1. Stimulus electrode to popliteal fossa electrode—divide this distance by the latency of the popliteal fossa potential.
2. Stimulus electrode to L3S electrode—divide this distance by the L3 latency.
3. Popliteal fossa electrode to L3S electrode—divide this distance by the difference in the latency of the popliteal fossa and L3 potentials.

For instance, if posterior tibial nerve stimulation produces a popliteal potential with a peak at 9 msec and an L3 potential with a latency of 21 msec, and if the distance between stimulus electrode and popliteal fossa electrode measures 40 cm and the distance between stimulus electrode and L3S electrode is 90 cm, the conduction velocity between stimulus electrode and popliteal fossa equals 400 mm divided by 9 msec, or 44 m/sec, and the conduction velocity between stimulus electrode and L3S equals 900 mm divided by 21 msec, or 43 m/sec. The conduction velocity between popliteal fossa and lumbar electrodes equals the distance between these electrodes, or 500 mm, divided by the latency difference of 12 msec, or 42 m/sec.

### 19.1.2 Slow Central Conduction Velocity

Central conduction velocity is determined by dividing the conduction distances between recording electrodes on the spinal cord and scalp by the difference in the peak latencies at these points, that is, by the central conduction times. The resulting values do not represent true conduction velocities because the measurements on the skin are somewhat longer than those of the spinal cord and because the travel between the recording points includes synaptic transmission in addition to axonal conduction. Besides, like peripheral conduction velocity, central conduction velocity is measured to the peak rather than to the onset of a deflection and therefore does not reflect the maximum speed of conduction. Furthermore, the electric potentials recorded at different points are composed of different fractions of presynaptic and postsynaptic elements and do not necessarily represent equivalent indicators of the passage of nerve impulses.

Three central conduction velocities can be calculated for common peroneal stimulation. Straight-line distances are measured from the L3S, T12S, and T6S electrodes to the Cz' electrode and are used as central conduction distances. These central conduction distances are divided by the corresponding central conduction times obtained by subtracting from the peak latency of the P27 the peak latencies of the L3, T12, and T6 potentials. Two central conduction velocities can be calculated for posterior tibial nerve stimulation by measuring the distances from the two spinal recording electrodes used with this stimulus, namely, L3S and T12S, to Cz' and dividing these distances by the latency differences between the L3 and P37 peaks and between the T12 and P37 peaks, respectively.

For instance, if posterior tibial nerve stimulation produces an L3 potential at a latency of 19 msec, a T12 potential at 21 msec, and a scalp SEP with a P37 at 37 msec, and if the distance from the L3S to Cz' measures 70 cm and that from T12S to Cz' is 60 cm, the central conduction velocity between L3S and Cz' equals 700 mm divided by 18 msec, or 39 m/sec, and the central conduction velocity between T12S and Cz' equals 600 mm divided by 16 msec, or 38 m/sec.

## 19.2 GENERAL CLINICAL INTERPRETATION OF ABNORMAL SEPS TO LEG STIMULATION

Table 19.1 shows the clinical interpretation of SEPs to leg stimulation. Table 19.2 shows the most common leg SEP findings in various disorders.

Absence of all SEPs, unless explained by technical problems, indicates a lesion at or below the cauda equina. A decrease of peripheral conduction velocity has the same significance when technical problems are excluded. Stimulation of the posterior tibial nerve affords recordings from two points over the nerve and therefore permits distinction between distal peripheral neuropathy and proximal peripheral nerve or plexus neuropathy. In the case of distal involvement, the popliteal fossa potential and all subsequent SEPs may be absent, or peripheral conduction velocity to the popliteal fossa electrode may be slowed, whereas in the case of more proximal lesions, SEPs at the lumbar electrode may be absent, or peripheral conduction to that electrode may be abnormally slow and popliteal fossa potentials may be normal. With either posterior tibial or common peroneal nerve stimulation, the abolition of spinal potentials above a normal L3 potential raises the suspicion of a high lumbar or low thoracic spinal lesion but can not prove such a lesion because cord potentials are highly variable in normal subjects and more difficult to obtain at higher levels. For the same reason, slowing of conduction between cord segments can not be taken as evidence for lesions between these segments. The absence of scalp SEPs or a slowing

TABLE 19.1. Clinical interpretation of SEPs to stimulation of the common peroneal nerve (CPN) at the knee or the posterior tibial nerve (PTN) at the the ankle

| Abnormality | Interpretation |
| --- | --- |
| A. Posterior tibial nerve stimulation | |
| 1. Absent PF ± absent L3 potential with normal scalp SEP | Normal |
| 2. Decreased peripheral NCV to PF with normal PF–L3S NCV | Lesion between ankle and PF |
| 3. Decreased peripheral NCV to PF and PF–L3S with normal CCV | Lesions of both distal and proximal peripheral nerves |
| 4. Decreased PF–L3S NCV with normal NCV to PF | Lesion between PF and cauda equina |
| 5. Absent L3 and T12 potentials with normal PF; scalp potential absent or delayed | Suspect lesion between PF and cauda equina |
| B. Common peroneal nerve stimulation | |
| 1. Absent L3 with present or absent T12 and T6 and normal scalp SEP | Normal |
| 2. Absent L3, T12, and T6 with delayed or absent scalp SEPs | Defect at or above the cauda equina, or both |
| 3. Decreased peripheral conduction velocity to L3 | Peripheral defect between PF and cauda equina |
| C. Either posterior tibial or common peroneal stimulation | |
| 1. Absent SEPs to leg stimulation at all levels | Peripheral nerve lesion; rule out technical problem |
| 2. Increased latency of SEPs at all recording levels | Hypothermia; inaccurate measurement of distances between stimulating and recording electrodes |
| 3. Decreased CCV | Defect above the cauda equina and below or at the somatosensory cortex |
| 4. Absent scalp SEP | Suspect defect above the cauda equina and at or below the somatosensory cortex |
| 5. Decreased peripheral NCV and decreased CCV | Lesions above and below the cauda equina, or a single lesion at the cauda equina or lower spinal cord |

of central conduction velocity, with preserved spinal potentials and normal peripheral conduction velocities, may be due to a lesion above the lumbar spinal cord, in the brainstem, or the cerebral hemisphere opposite the stimulated leg; however, absent scalp SEPs may be due to technical problems with the recording of these often very small potentials.

## 19.3 PERIPHERAL NERVE AND ROOT LESIONS THAT CAUSE ABNORMAL SEPS TO LEG STIMULATION

### 19.3.1 Peripheral Nerve Lesions

Peripheral nerve lesions are most conveniently diagnosed by a decreased peripheral conduction

TABLE 19.2. SEPs to leg stimulation in various disorders

| Disorder | Finding |
| --- | --- |
| Adrenoleukodystrophy | Increased CCT |
| Charcot-Marie-Tooth disease | Normal |
| Diabetes mellitis | Peripheral slowing and occasional central slowing |
| Freidreich's ataxia | Increased CCT |
| Multiple sclerosis | Increased lumbar scalp CT; normal peripheral CT |
| Myotonic dystrophy | Increased peripheral CT |
| Parasagittal cerebral lesions | Abolish cortical potentials |
| Radiculopathy | Usually normal |
| Spinal cord injury | Often abnormal; early return or normal response indicates favorable prognosis |
| Subacute combined degeneration | Cortical potentials delayed or abolished |

CT = conduction time
CCT = central conduction time

velocity. However, scalp SEPs may be easier to record than the diminished distal SEPs and have been used to evaluate peripheral conduction.

### 19.3.2 Radiculopathy

Nerve root compression has been found to not alter SEPs reliably. Stimulation of several cutaneous nerves or of dermatomes may be needed for a clear identification of the involved roots, but conventional EMG probably has greater diagnostic power, especially in cases with motor deficits.[57] One study suggested that SEPs were more sensitive than EMG for diagnosis of radiculopathy, but were less helpful in the diagnosis of spinal stenosis.[89]

### 19.4 LESIONS OF SPINAL CORD AND BRAINSTEM THAT CAUSE ABNORMAL SEPS TO LEG STIMULATION

### 19.4.1 Multiple Sclerosis

Leg SEPs are often abnormal due to lesions at spinal or supraspinal levels, which may cause increased latency of the scalp SEP with a normal lumbar SEP, that is, an increased lumbar-scalp conduction time and decreased central conduction velocity (Figures 19.1 and 19.2). The scalp SEP may be entirely abolished. SEPs to leg stimulation are more often abnormal in MS than are SEPs to arm stimulation or other EPs, probably because they test a longer pathway which is more likely to be affected by the scattered lesions of MS.

### 19.4.2 Spinal Cord Injury

Complete interruption of ascending spinal pathways abolishes SEPs above the lesion. Incomplete cord lesions, especially those reducing joint position sense, abolish or delay SEPs. SEPs both elicited and recorded either below or above the lesion may be preserved. For instance, the lumbar SEP to leg stimulation and SEPs to arm stimulation may be present in lesions above the cauda equina and below the cervical spinal cord. However, in some cases the SEP is abnormal below the level of a lesion.

Patients with spinal cord injury may have a different relationship between latency and inten-

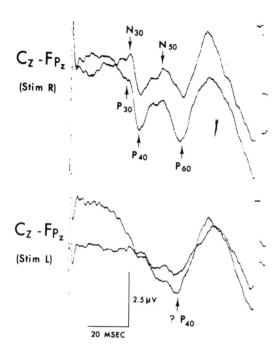

FIGURE 19.1. *Scalp SEPs to stimulation of the posterior tibial nerve in a patient with possible MS. Stimulation of the right tibial nerve produces a normal SEP (top tracings). Stimulation of the left tibial nerve produces a SEP showing an absent or delayed first positive peak (bottom tracing). Two tracings are superimposed for each SEP. Negativity at Cz is plotted upward. This 54-year-old woman has progressive spinal MS. From Eisen and Odusote (EEG Clin Neurophysiol 48:253, 1980) with permission of the authors and Elsevier Scientific Publishers Ltd.*

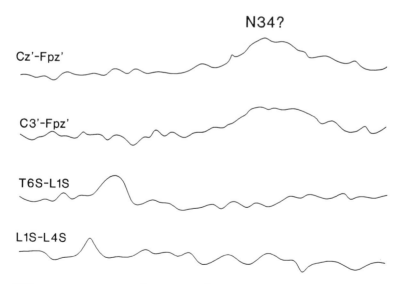

FIGURE 19.2. *SEPs to common peroneal nerve stimulation in a patient with possible MS. Stimulation of the left common peroneal nerve (arrow) elicits a negative peak of over 10 msec in the lumbar recording (bottom tracing) and of about 15 msec in the lumbothoracic recording (third tracing). Scalp recordings show only a late negative wave, with a peak at over 40 msec (top two tracings). Stimulation and recording methods as in Figure 18.4. This 16-year-old boy developed progressive weakness and numbness of both legs and urinary retention. He had hyperreflexia of both legs, slightly decreased touch and pain sensation up to the level of L3, and markedly decreased vibration and position sense. Myelogram and CSF were normal.*

sity, such that increasing intensity produces more of a reduction in latency than observed in normals.[7]

SEPs may recover before the clinical condition improves. Presence of an SEP soon after injury, early return, and progressive normalization of the SEP waveform usually indicate a favorable prognosis.

### 19.4.3 Spinal Cord Compression

Compression of the spinal cord by cervical spondylosis, extramedullary and intramedullary tumors, and Hodgkin's disease may reduce or delay the cortical SEP if these lesions interfere with position sense, but the SEP remains intact in many cases of extramedullary spinal lesions.

In the Brown-Séquard syndrome, the SEP is abnormal on stimulation of the side with decreased vibration and position sense, but not on stimulation of the other side. The abnormal SEP may persist after removal of the lesion and recovery of normal sensory function.

### 19.4.4 Charcot-Marie-Tooth Disease, Friedreich's Ataxia, Olivopontocerebellar Degeneration, Adie's Syndrome, and Tabes Dorsalis

Scalp SEPs are often severely abnormal. Central conduction times have been found to be normal in peroneal muscular atrophy but increased in Friedreich's ataxia.

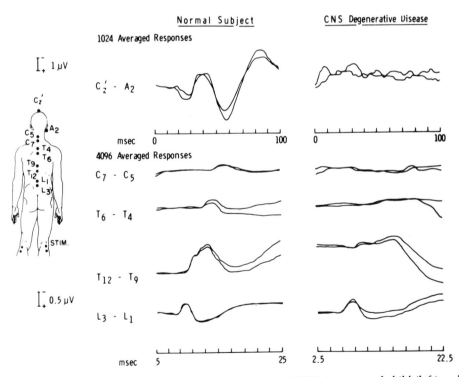

FIGURE 19.3. *Spinal (bottom four tracings) and scalp (top tracing) SEPs in a normal child (left) and a child with degenerative CNS disease (right). In the normal child, SEPs disappear at rostral spinal levels, and no scalp SEP is recorded. Bilateral peroneal nerve stimulation at the knee. Negativity at the first electrode named in each pair at the left of the tracings is plotted upward. SEPs in the patient have shorter latency because of a shorter recording distance. From Cracco et al. (EEG Clin Neurophysiol 49:437, 1980) with permission of the authors and Elsevier Scientific Publishers Ireland Ltd.*

### 19.4.5 Myotonic Dystrophy

Increases of SEP latency suggest motor, sensory, or mixed nerve involvement in many cases.

### 19.4.6 Subacute Combined Degeneration of the Spinal Cord

SEPs are commonly abnormal. They show significant delays or are abolished with marked myelopathy and return after treatment.

### 19.4.7 Diabetes Mellitus

Diabetics with mild or no clinical signs of polyneuropathy may show slowing of conduction not only in peripheral nerves, but also in the spinal cord, suggesting subclinical dysfunction of the posterior columns.

### 19.4.8 Subacute Myelo-Opticoneuropathy

Clioquinol intoxication, reported from Japan, leads to widespread CNS involvement, including marked posterior column damage; this affects central conduction of leg SEPs.

### 19.4.9 Degenerative CNS Diseases in Children

Spinal cord involvement has been reported to produce abnormal SEPs in children with various degenerative diseases (Figure 19.3). Scalp SEPs are often delayed in adrenoleukodystrophy and Friedreich's ataxia.

### 19.4.10 Surgical Monitoring of Spinal Cord Condition

Compression, traction, and ischemia of the spinal cord during surgery may impair conduction through sensory pathways and cause SEP changes even before producing lasting damage. The scalp SEP to leg stimulation may therefore be used as a monitor of the condition of the spinal cord during spinal surgery. A deterioration of the scalp SEP can indicate impending cord damage if other causes, such as changes in the level of anesthesia, of blood pressure, and of ventilation, are excluded.[91] Recordings from electrodes placed into the epidural space, inserted into the interspinal ligament, or implanted into the spinal process above the operative site have also been used for monitoring and are less susceptible to changes of blood pressure and anesthetic levels than scalp recordings.

## 19.5 CEREBRAL LESIONS THAT CAUSE ABNORMAL SEPS TO LEG STIMULATION

Parasaggital tumors have been reported to abolish the $\overline{N70}$ of the scalp SEP to leg stimulation. The amplitude of the scalp SEP to leg stimulation is reduced in patients with Huntington's disease and in many subjects at risk for this disease.

# E

# Event-Related and Other Potentials

---

## PART CONTENTS

# 20

# Other Sensory Evoked Potentials

## 20.1 SENSORY NERVE ACTION POTENTIALS

Sensory nerve action potentials (SNAPs) are not used routinely with SEP recordings but may be studied in cases requiring investigation of sensory nerve function (Figure 17.2). In these cases, stimulus electrodes are placed over sensory nerves or sensory branches of mixed nerves, and recording electrodes are placed proximally over the same nerve. The amplitude of SNAPs is usually low even after averaging large numbers of responses and may be increased by recording with fine needle electrodes inserted subcutaneously close to the nerve.

SNAPs have a major negative peak, usually preceded and followed by minor positive peaks, with a latency depending on conduction distance and fiber speed. This may be followed by later peaks representing fibers of lower conduction speed. The latency of SNAPs may decrease with increasing stimulus intensity, presumably because of wider spread of stimulus current along the stimulated nerve. Sensory conduction velocity is calculated by dividing the distance from the stimulating to the recording electrode by the latency of the negative SNAP peak. Conduction in sensory nerve fibers can also be measured during recording of arm or leg SEPs if sensory nerve or nerve branches are used for stimulation. Recordings from clavicular and lumbosacral electrodes then represent SNAPs.

## 20.2 SEPS TO TRIGEMINAL NERVE STIMULATION

Electric stimulation of the lips, gums, or mental nerve produces scalp SEPs consisting of a series of peaks that probably represent the sensory pathway from Gasser's ganglion to the cerebral cortex (Figures 20.1 and 20.2). Trigeminal SEPs may be abnormal in MS, trigeminal neuralgia (Figure 20.2), brainstem infarcts and tumors, after Gasserian thermocoagulation leading to hypesthesia, in acoustic neurinoma, and in sarcoidosis involving the trigeminal nerve.

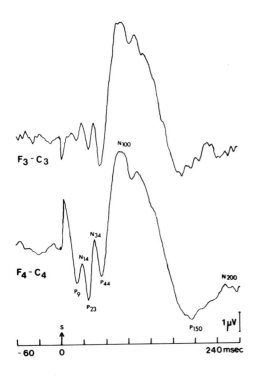

FIGURE 20.1. *Normal trigeminal SEPs. Stimulation of the left trigeminal nerve produces a complex left (top tracing) and right (bottom tracing) scalp SEP with several deflections representing subcortical and cortical peaks. Stimulation of the third trigeminal branch at the mental foramen. Recordings between frontal and central electrodes. Negativity at the frontal electrode is plotted upward. From Drechsler (In Evoked potentials: Proceedings of an international evoked potentials symposium held in Nottingham, England, ed. C Barber, pp. 415–422) with permission of the author and MTP Press Ltd.*

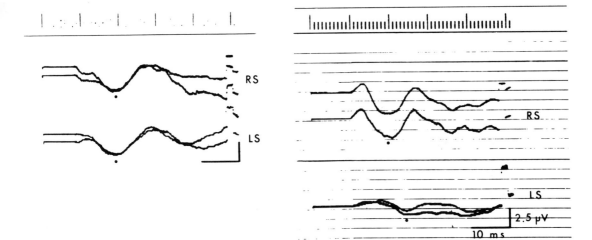

FIGURE 20.2. *Trigeminal SEPs in a normal subject (left) and a patient with trigeminal neuralgia (right). In the normal subject, SEPs from the right side (RS) and left side (LS) of the scalp to stimulation of the contralateral lips show a positive peak at about 20 msec. In the patient with trigeminal neuralgia, stimulation of the intact left side produces a normal SEP over the right side of the head (RS), whereas stimulation of the affected right side produces a SEP with a peak of lower amplitude and longer latency over the left side (LS). Two tracings are shown for each SEP. Simultaneous stimulation of upper and lower lips opposite the recording site. Recordings between C5 or C6 and an Fz reference electrode of the International 10–20 System. Negativity at the central electrodes is plotted upward. From Stöhr et al. (Ann Neurol 9:63, 1981) with permission of the authors and Little, Brown and Company.*

## 20.3 SEPS TO PUDENDAL AND BLADDER STIMULATION

SEPs to stimulation of mucous membrane in the distribution of the pudendal nerve and inside the bladder have been explored in normal subjects with the expectation that these SEPs may become useful in the evaluation of patients with sexual, bladder, or bowel dysfunctions.

## 20.4 SEPS TO PHYSIOLOGICAL STIMULI

In contrast to the routine use of adequate stimuli for VEPs and AEPs, such stimuli have been used only rarely for SEPs. Because SEPs are mediated through the lemniscal system, touch is the most effective adequate sensory stimulus. Even though recordings have been made in a few neurological disorders, no routine diagnostic tests have yet been developed.

### 20.4.1 Touch

Sudden displacement of the skin or of a fingernail elicits clavicular, cervical, and scalp SEPs similar to those produced by electrical stimulation of the corresponding nerve except that the peaks are generally less clearly defined. Short stimuli of 5 msec are as effective as longer ones, and stimulus rates of 8/sec and more may be used. Latency varies with stimulus site and is very short for the tongue. Stimulation of the trunk and the proximal parts of arms and legs tends to give bilateral responses. SEPs to mechanical stimulation may be abnormal in patients with peripheral nerve and central sensory lesions but are normal in persons with hypnotically induced anesthesia.

### 20.4.2 Vibrotactile Stimuli

Vibratory stimulation of the skin is less suitable for producing discrete short-latency SEPs than is an intermittent touch stimulus, but the effect of vibratory stimuli can be studied by their interference with SEPs to concurrent electric nerve stimuli.

### 20.4.3 Joint Movement

Passive displacement of fingers produces scalp SEPs with latencies comparable to those of electrically induced SEPs and probably mediated by joint capsule afferents. Active and passive joint movements modulate SEPs to electric stimulation.

### 20.4.4 Muscle Stretch

Sudden muscle stretch elicits scalp SEPs of short latency, probably mediated by proprioceptive afferents.

### 20.4.5 Pain Stimuli

Painful radiant heat to the skin, or electric stimuli to peripheral nerves produce SEPs different from those of painless stimuli. However, electric stimulation of the skin can probably not excite C-fibers in numbers sufficient for an SEP without producing tissue damage. Electric shocks to the tooth pulp, generating a pure pain stimulus, cause long-latency cortical SEPs that do not correlate precisely with perceived pain.

### 20.4.6 Warm and Cold Stimuli

Changes in skin temperature, although difficult to control precisely, produce SEPs.

## 20.5 SOMATOMOTOR EPS

Like click stimuli, electric stimulation of mixed or sensory nerves can induce muscle responses that may begin about 15 msec after the stimulus, especially in contracted muscles, and may contaminate recordings of cerebral and spinal SEPs from scalp and neck. These potentials can often be distinguished from SEPs by their disappearance with relaxation and by their greater prominence at the periphery of the scalp near larger muscle groups.

# 21

# Event-Related Potentials, Olfactory Evoked Potentials, and Magnetic Evoked Potentials

## 21.1 THE P$\overline{300}$

### 21.1.1 Methods of Producing the P$\overline{300}$

A P$\overline{300}$ may be elicited in various ways. Usually, two kinds of stimuli are mixed in one sequence, and the subject is asked to pay attention to one kind, usually the rarer one, and to ignore the other. This is sometimes referred to as the *odd-ball paradigm* and often implemented with tones of high and low frequency ("beep-boop" stimuli). Responses to each kind of stimulus are recorded between an electrode over the central or posterior head regions and a distant reference electrode and are averaged separately. A P$\overline{300}$ appears after the AEP to stimulation with the rare stimulus if the subject pays attention to this stimulus (Figure 21.1). These methods require the facility to present two kinds of stimuli, mixed irregularly in different proportions, and to sort the responses to each kind of stimulus in separate averaging channels.

A P$\overline{300}$ may also be obtained by instructing the subject to guess whether the next stimulus will be visual or auditory or whether it will be a single or double click. The subject may be asked to count or otherwise pay attention to brief presentations of certain visual patterns or colors, words, letters, or sounds that occur irregularly and infrequently within a series of visual or auditory stimuli from which they differ in some regard. Even the omission of a stimulus in a train of uniform and regular stimuli may produce a P$\overline{300}$ at a fixed time after the expected stimulus, clearly indicating that this potential is not evoked by an exogenous stimulus but generated endogenously. On the other hand, the simple presentation of an unexpected stimulus unrelated to any instructions may also be followed by a P$\overline{300}$ which, however, has a shorter latency and a more anterior distribution. Thus, although the production of the P$\overline{300}$ seems to depend on cognitive factors, the precise physiological mechanisms involved are unclear.

### 21.1.2 The Normal P$\overline{300}$

The P$\overline{300}$ is a positive peak with a latency of 250–600 msec (Figure 21.1). It may be preceded

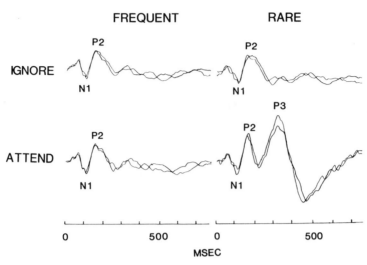

FIGURE 21.1.    *The normal P300. Frequent (left) and rare (right) auditory stimuli are intermixed irregularly and the responses are averaged separately. When both stimuli are ignored (top row), they produce similar EPs with N1 and P2 peaks (top left and right). When the rare stimuli are attended to (bottom row), frequent stimuli (bottom left) produce the same peaks as do ignored stimuli, but rare stimuli (bottom right) are followed by an additional positive peak, the P300 (P3). From Squires and Hecox (Semin Hear 4:415, 1983) with permission by the authors and Thieme-Stratton Inc.*

by EP peaks if the recording electrodes are located near the cortical receiving area for the sensory modality of the stimulus used. The latency of the P300 increases with the time the subject needs to distinguish the rare stimulus; the amplitude increases with the rarity of the stimulus and, to some extent, with stimulus intensity. Maneuvers affecting the P300, such as manipulations of the subject's attention, may also alter the amplitude of the preceding long-latency peaks of VEPs, AEPs, and SEPs.

### 21.1.2.1 Distribution

The P300 has a wide distribution with a maximum in the parietotemporal areas, unrelated to the specific sensory areas but probably related to the parietotemporal association cortex and to subcortical structures such as the hippocampus or thalamus. Even though the P300 to novel stimuli may be located more frontally, the distribution of the P300 generally does not reach as far forward as that of the contingent negative variation.

### 21.1.2.2 Effect of age

A P300 to unfamiliar stimuli has been obtained in infants of 3 months. Word stimuli normally become effective at 8–12 years. Between childhood and adulthood, the distribution of the P300 becomes more restricted to the parietal area and the latency decreases. In adults, aging increases the latency, decreases the amplitude, and causes a forward shift in the distribution of the P300.

### 21.1.3 The Abnormal P300

Although the latency of the P300 varies widely in normal subjects, it has been reported to be prolonged in adult patients with mental retardation, in patients with dementia of various causes, in patients with chronic renal failure, in patients with Parkinson's disease, in chronic alcoholics shortly after detoxification, and in schizophrenic patients. Patients with temporal lobe epilepsy may have a delayed P300 which is independent of seizure manifestation or antiepileptic drugs.[37] Abnormalities in P300 latency may develop early in Alzheimer's disease and progress proportionally to intellectual decline.[5,68]

Delayed P300 has been reported in patients with AIDS and AIDS-related complex even before psychometrics could detect cognitive deficits.[62] Children who received both prophylactic cranial irradiation  intrathecal methotrexate for acute lymphocytic leukemia had delayed P300.[74]

Changes of P300 amplitude have been described in various conditions, including frontal lobe lesions, hyperactive children treated with methylphenidate, increased blood lead levels in children, infantile autism, and schizophrenia, but these changes are of uncertain diagnostic importance because of the great normal variability of the P300 amplitude.

## 21.2 THE CONTINGENT NEGATIVE VARIATION (CNV)

### 21.2.1 Methods of Producing the CNV

The CNV is usually produced by presenting pairs of stimuli separated by 1–2 sec; the first stimulus serves as a warning signal and the second, or imperative, stimulus as a signal requiring a response. A common example is a click followed after about 1 second by a series of flashes to which the subject is required to respond by pushing a button. Repetition of the same sequence leads to the buildup of a negative potential that begins after the EP to the first stimulus and ends with the EP to the second stimulus (Figure 21.2). The modality and order of presentation of the stimulus are unimportant: A flash followed by clicks requiring a response is equally effective. However, the mere presentation of paired stimuli without the requirement of a response to the second stimulus does not produce a CNV. Moreover, the amplitude of the CNV decreases if the warning stimulus is not always followed by the imperative stimulus; the

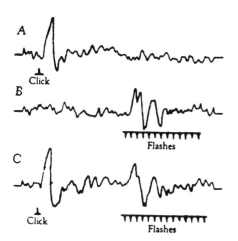

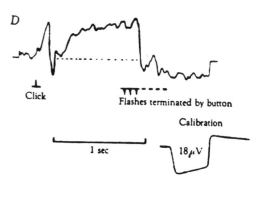

FIGURE 21.2.   *The normal CNV. (A) Stimulation with a click produces a LLAEP. (B) Stimulation with a train of flashes elicits a VEP. (C) Click stimuli followed by flash stimulation produces a sequence of the EPs shown in (A) and (B). (D) Click stimulation followed by flash stimulation gradually leads to the development of a slow negative wave between the click and flash EPs when the subject follows the instruction to press a button terminating the flashes. Recording between vertex and mastoid electrodes. Negativity at the vertex is plotted upward. From Walter et al. (Nature 203:380, Copyright (c) 1964 Macmillan Journals Ltd.) with permission of the authors and the publisher.*

CNV usually disappears entirely if less than 70% of warning stimuli are followed by imperative stimuli.

If two different warning stimuli are used and the subject is asked to respond to only one of them, a CNV will develop only after that stimulus. This paradigm may be used to test the ability of the subject to discriminate between similar stimuli and to test the effectiveness of the first stimulus. If the intensity of the warning stimulus is reduced below the perceptual threshold, it becomes ineffective in generating a CNV.

The CNV is recorded between electrodes on vertex and ear or mastoid. The averaging epoch should be 2–4 seconds long to permit identification of a baseline from which the CNV amplitude can be measured. The CNV is enhanced by averaging 10 or more responses. The technical requirements for eliciting and recording the CNV go beyond those of routine EP recordings. Stimulation requires the ability to generate paired stimuli, usually of different modality. Recording facilities must be able to register very slow, long-lasting electric potential changes and to verify the subject's response to the second stimulus.

## 21.2.2 The Normal CNV

The CNV begins about 400 msec after the warning stimulus and forms a ramp or rectangle having a maximum within about 800 msec of the first stimulus; it has an amplitude of usually up to 50 μV (Figure 21.2). The CNV develops gradually as the paired stimuli are repeated. The speed of development differs among subjects. Anxiety and distraction reduce the CNV amplitude. Because a CNV does not develop if the subject fails to perceive the warning stimulus or to respond to the imperative stimulus, a CNV cannot be elicited in infants or uncooperative subjects.

An increase of the interval between the warning and the imperative stimulus to about 4 sec splits the CNV into two waves that have different distribution and functional characteristics, suggesting that the CNV may consist of a combination of long-latency potentials evoked by the first stimulus and of potentials

related to the preparation for a reaction to the second stimulus.

The name *contingent negative variation* has been applied because this potential is contingent on a stimulus-response sequence, is electrically negative, and is a slow variation from the electric baseline. Because it seems to depend on the subject's expectation of the second stimulus, it has also been called *expectancy wave*. The methods of producing a CNV suggest that the CNV depends on attention, motivation, effort, preparation for action, and other psychological variables.

### 21.2.2.1 Distribution

The CNV has a maximum at the vertex. It extends farther frontally than parietally and has a minimum in the occipital and posterior temporal areas.

### 21.2.2.2 Effect of age

In old age, the CNV has been reported to increase and to depend on the subject's ability to switch attention.

### 21.2.3 The Abnormal CNV

Only a few studies have attempted to relate the CNV to neurological and psychiatric diseases. The CNV has been used in audiological studies to determine the threshold of auditory perception or the discrimination of phonemes.

## 21.3 READINESS POTENTIAL (BEREITSCHAFTSPOTENTIAL) AND OTHER MOVEMENT-RELATED POTENTIALS

Computer averaging of the electric activity that precedes a voluntary movement shows a gradually rising negative potential that begins about 1 sec before the movement, has a wide, bilateral distribution with a maximum at the vertex and probably reflects the process of getting ready to move; no such potentials are seen before passive movement (Figure 21.3). Both passive and active movements are followed by several peaks.

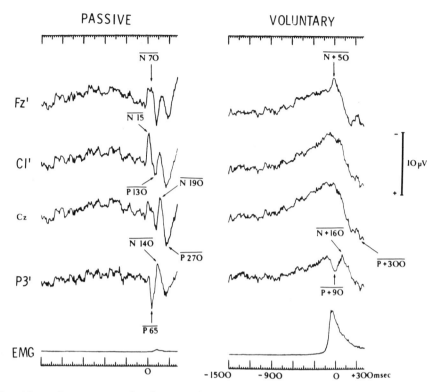

**FIGURE 21.3.** *Normal movement-related potentials. Passive extension of the right middle finger (left half, at time 0) is followed by sequences of peaks having different latencies in various locations (top to bottom). Volitional right middle finger extension (right half, at time 0) is preceded by a widespread, slowly increasing negative deflection and followed by simpler fast waves. Each tracing is the grand average of recordings from four subjects. Recordings from midfrontal, left central, vertex, and left parietal electrodes in reference to linked ear electrodes. Negativity at the scalp electrodes is plotted upward. From Shibasaki et al. (EEG Clin Neurophysiol 50:201, 1980) with permission of the authors and Elsevier Biomedical Press B.V.*

Movement-related potentials have now been described in great detail. Only a few attempts have so far been made to use these potentials for the study of cerebral disorders, for instance, Parkinson's disease, cerebellar disorders, and Tourette's syndrome.

## 21.4 POTENTIALS PRECEDING SPEECH AND ACCOMPANYING WRITING

Slow negative potentials can be recorded several seconds before the onset of speech. They have a frontal distribution with an asymmetry suggesting a specific relation to speech centers. Poten-

tials with a maximum at the vertex and the left motor area appear during writing in right-handed subjects.

## 21.5 OLFACTORY EPS

Olfactory stimuli elicit EPs of relatively long latency and duration. Ensuring precise stimulus and recording conditions poses many problems. Even when they are overcome, it is difficult to be certain that EPs to odorous stimuli are due to specific olfactory stimulation rather than to excitation of the nasal somatosensory fibers of the trigeminal nerve.

## 21.6 MAGNETIC EVOKED FIELDS

Visual, auditory, somatosensory, and cognitive stimuli and finger, foot, and toe movements induce changes of the magnetic field about the head that can be recorded and averaged. These magnetic evoked fields generally resemble the corresponding EPs and ERPs and may give some information regarding the localization of the generators in the head.

An array of detectors is placed on the head to detect fluctuations in magnetic fields in response to the stimulus or movement. The potentials can be reconstructed by computer to form a map of the magnetic fields.

Somatosensory-evoked magnetic fields may show abnormal patterns in cerebral disordesr, such as MS,[45] but these studies are still not used in routine clinical practice.

# 22

# Motor Evoked Potentials

Motor evoked potentials are not routinely used by most neurophysiologists for several reasons. Special equipment is required, which is not expensive, but is usually not supplied with an evoked potential machine at time of purchase. Many physicians are afraid of liability associated with electrical stimulation of the brain, even though the risk of injury or kindling seizures is miniscule. Of course, there is a poor correlation between perceived medicolegal liability and merit of the case. The biggest limitation to the use of motor evoked potentials is the clinical utility. New imaging techniques and sensory evoked potentials can provide most of the information that motor evoked potentials offer.

## 22.1 PHYSIOLOGICAL BASIS

Motor evoked potentials are the responses of muscles to stimulation of the cerebral cortex and spinal cord. Because recordings are made only from the muscles, motor conduction velocities between cortex and spinal cord can be cal-

culated. This is directly analogous to peripheral motor nerve conduction velocities.

Delayed conduction of central motor axons can be produced by spinal cord lesions, where there is compression or otherwise impaired action potential propagation. Structural lesions of the brain affecting descending motor axons and motoneuron degeneration can also produce abnormal motor conduction.

## 22.2 METHODS

### 22.2.1 Stimulus

#### 22.2.1.1 Electrical stimulation

The stimulus can be electrical or magnetic. Magnetic stimulation is preferable. Electrical stimulation is anodal, as opposed to cathodal stimulation used for peripheral nerve stimulation. The most commonly employed technique uses a flat anode placed over the motor strip and a long beltlike cathode that goes around the head. Recording electrodes are placed over the muscle to be studied, often the abductor digiti minimi. The patient is asked to make a mild voluntary contraction of the muscle. This facilitates the response so that lower amplitude stimulation is required. Therefore, the field of anodal stimulation is very restricted. This prevents cortical stimulation from producing extensive muscle contraction.

The spinal cord can be stimulated near the C6 or T12 spinous processes. These are located to stimulate the cervical and lumbar enlargements, respectively. Voluntary activation does not facilitate the response to spinal cord stimulation.

### 22.2.1.2 Magnetic stimulation

Magnetic stimulation is provided by a hand-held coil of wire encased in plastic. Pulses of current are delivered through the wire. Charge movement through the coil creates a magnetic field like an inductor. The magnetic field causes charge movement in cortical neurons. The cell bodies are depolarized and reach threshold.

Magnetic stimulation has advantages over electrical stimulation:

- Magnetic stimulation is not perceived as painful.
- The skin does not have to be prepared, as it does for electrical stimulation.
- Neurophysiologists are not so concerned about the risks of magnetic stimulation as of electrical stimulation, even though the electrophysiological result of both modalities is the same.

### 22.2.2 Recording

There are no universally agreed-upon guidelines for performing motor evoked potential studies. These were derived from King and Chiappa (1990) in the book entitled *Evoked Potentials in Clinical Medicine.*[47]

Low-frequency filter is set to 3 Hz and high-frequency filter to 3,000 Hz. Gain is determined by the amplitude of the response. Set the gain so that the response produces a deflection that is at least 50% of the maximal excursion, but does not go off scale. Time base is set initially at 5 ms/div.

Recordings are made from arm and occasionally leg muscles. The following upper extremity muscles are most commonly examined: first dorsal interosseus, abductor digiti minimi, abductor pollicis brevis, biceps, and triceps. In the lower extremity, the tibialis anterior is most often studied. The active recording electrode is placed over the belly of the muscle. The reference is placed distally, over the tendon or distal joint.

### 22.2.3 Testing Protocol

#### 22.2.3.1 Threshold determination

The stimulus intensity is gradually increased until a CMAP is seen. Then, the stimulus intensity is lowered to below threshold and successive stimuli given at regular incrementing intensities, for example, 5% increase per stimulus. In this way, the threshold is clearly defined. For magnetic stimulation, the threshold is represented as a percentage of the maximum output of the coil.

#### 22.2.3.2 Motor evoked potential recording

*22.2.3.2.1 Cortical stimulation.* The study is usually performed while the subject is at rest and with facilitation. Stimulus intensity is set at 15% above threshold, and three stimuli are delivered. The traces should superimpose. If they fail to superimpose, ensure that stimulating and recording electrodes are secure, and that the patient is completely relaxed.

Facilitation increases the size of the response while reducing latency. Many investigators believe a facilitated response to be more reproducible and therefore clinically useful than a resting response. Facilitation is achieved by asking the patient to make a submaximal contraction of the muscle. The contraction should be about 10% of the maximal voluntary contraction of the muscle.

*22.2.3.2.2 Spinal stimulation.* Spinal stimuli are delivered at the level appropriate for the muscle being tested. For electrical stimulation, the stimulating cathode is just above or below the C7 spinous process. Best stimulation is obtained with the cathode in the space between spinous processes. For magnetic stimuli the muscles and levels are:

C3: biceps
C4: triceps
C5: abductor minimi

Voluntary contraction produces little facilitation of the spinal response, so it is not used.

*22.2.3.2.3 Peripheral stimulation.* The site of stimulation again depends on the muscle being studied. For biceps and triceps, magnetic stimuli are delivered to Erb's point. Electrical stimulation at Erb's point is much more painful, and should not be used unless necessary. For hand muscles, electrical stimulation is delivered to the innervating nerves near the wrist.

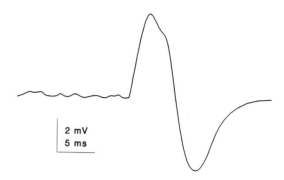

FIGURE 22.1. *Sample recording of a motor evoked potential. Magnetic stimulation of the cortex and recording from the abductor digiti minimi.*

### 22.2.4 Contraindications

Motor evoked potentials should probably not be performed if the patient has:

• skull defect, such as a burr hole or craniotomy defect
• metal in the head or neck
• history of epilepsy or photoconvulsive discharge on EEG
• less than 18 years of age

Special caution should be employed in studying patients with metal or electronic devices elsewhere in their bodies. Devices of special concern include pacemakers, implanted nerve stimulators, implanted pumps (magnetic field could change programming).

## 22.3 INTERPRETATION

### 22.3.1 Measurements

Latency and amplitude are measured for each of the waves recorded using cortical, spinal, and peripheral stimulation. Latency is from the time of stimulus until the first deflection of the wave. Latency to peak of the response may be measured but is not used for interpretation. Amplitude is measured from baseline to the peak of the response.

### 22.3.2 Normal Data

Motor evoked potentials are not in routine clinical use, so there are no standardized normative data. Figure 22.1 shows a sample recording. Central motor conduction time (CMCT) is the difference between cortical latency and spinal latency. Amplitudes are recorded, but are so variable that they are not used extensively in interpretation.

### 22.3.3 Abnormal Responses

A response is abnormal if it is absent or if the CMCT is more than 2.5–3 standard deviations from the mean. Abnormalities may be seen from one or both sides. The CMCT is used for interpretation rather than absolute response latency, so that conduction in the peripheral pathways is not considered.

Increased CMCT is seen in a variety of disorders. Some of these are:

• multiple sclerosis
• spinal cord lesions
• motor neuron disease
• Guillain-Barré syndrome
• hereditary spastic paraparesis

- stroke
- coma

Multiple sclerosis causes prolonged CMCT in the majority of patients. Some patients may have only reduced amplitude. Suspected multiple sclerosis is probably the chief indication for motor evoked potentials, as an adjunct to VEP, SEP, and MRI.

Intraoperative monitoring is probably going to be the main use of motor evoked potentials. SEP monitoring assesses the dorsal columns, supplied by the posterior spinal arteries. Motor evoked potentials will assess conduction in tracts supplied by the anterior spinal artery. The two modalities will likely be used together.

Spinal cord disorders frequently produce absence of the response or prolonged CMCT. Traumatic cervical myelopathy is almost always associated with abnormalities.

Motor neuron disease is characterized by absent responses. If present, the CMCT is only mildly prolonged. Some patients will have normal responses.

## 22.4 PITFALLS OF MOTOR EVOKED POTENTIALS

Technical pitfalls are similar to those described for sensory conduction studies. However, there are some pitfalls specific to motor evoked potentials.

Facilitation of the response to cortical stimulation requires that the patient make a voluntary contraction. Therefore, the muscle is active even prior to the stimulus. This can make determination of latency difficult. It can be hard to localize precisely the first deflection from baseline.

# REFERENCES

1. Allison T, Wood CC, Goff WR. 1983. Brainstem auditory, pattern-reversal visual, and short-latency somatosensory evoked potentials: Latencies in relation to age, sex, and brain and body size. Electroencephalogr Clin Neurophysiol 55: 619–636.

2. American Electroencephalographic Society Guidelines in EEG and evoked potentials, 1986. J Clin Neurophysiol 3(Suppl 1):43–92.

3. Apkarian P, Koetsveld-Baart JC, Barth PG. 1993. Visual evoked potential characteristics and early diagnosis of Pelizaeus-Merzbacher disease. Arch Neurol 50:981–985.

4. Arruga J, Feldon SE, Hoyt WF, Aminoff MJ. 1980. Monocularly and binocularly evoked visual responses to patterned half-field stimulation. J Neurol Sci 46:281–90.

5. Ball SS, Marsh JT, Schubarth G, Brown WS, Strandburg R. 1989. Longitudinal P300 latency changes in Alzheimer's disease. J Gerontol 44:195–200.

6. Bartel DR, Markland ON, Kolar OJ. 1983. The diagnosis and classification of multiple sclerosis: Evoked responses and spinal fluid electrophoresis. Neurology 33:611–17.

7. Beric A. 1992. Cortical somatosensory evoked potentials in spinal cord injury patients. J Neurol Sci 107:50–59.

8. Bird TD, Griep E. 1981. Pattern reversal visual evoked potentials. Studies of Charcot-Marie-Tooth hereditary neuropathy. Arch Neurol 38:739–41.

9. Blom JL, Barth PG, Visser SL. 1980. The visual evoked potential in the first six years of life. EEG Clin Neurophysiol 48:395–405.

10. Blumhardt LD, Barrett G, Kriss A, Halliday AM. 1982. The pattern-evoked potential in lesions of the posterior visual pathways. Ann NY Acad Sci 388:264–89.

11. Blumhardt LD, Barrett G, Halliday AM. 1977. The asymmetrical visual evoked potential to pattern reversal in one half-field and its significance for the analysis of visual field defects. Br J Ophthalmol 61:454–61.

12. Bodis-Wollner I, Hendley CD, Mylin LH, Thornton J. 1979. Visual evoked potentials and the visuogram in multiple sclerosis. Ann Neurol 5:40–47.

13. Boston JR. 1981. Spectra of auditory brainstem responses and spontaneous EEG. IEEE Trans Biomed Eng 28:334–341.

14. Buchthal A, Belopavlovic M. 1992. Somatosensory evoked potentials in cerebral aneurysm surgery. Eur J Anes 9:493–97.

15. Butinar D, Trontelj JV, Khuraibet AJ, Khan RA, Hussein JM, Shakir RA. 1990. Brainstem auditory evoked potentials in Wilson's disease. J Neurol Sci 95:1163–69.

16. Cacace AT, Shy M, Satya-Murti S. 1980. Brainstem auditory evoked potentials: A comparison of two high-frequency filter settings. Neurology 30:765–67.

17. Campbell JA, Leandri M. 1984. The effect of high pass filters on computer-reconstructed evoked potentials. EEG Clin Neurophysiol 57:99–101.

18. Chatrian GE, Wirch AL, Lettich E, Turella G, Snyder JM. 1982. Click-evoked human electrocochleogram: Noninvasive recording method, origin, and physiologic significance. Am J EEG Tech 22:151–174.

19. Chiappa KH. 1992. Short-latency somatosensory evoked potentials: Interpretation. In Evoked potentials in clinical medicine, 2nd ed. Chap. 7, ed. KH Chiappa, Raven, New York.

20. Chiappa KH, Harrison JL, Brooks EB, Young RR. 1980. Brainstem auditory evoked responses in 200 patients with multiple sclerosis. Ann Neurol 7:135–43.

21. Cohen SN, Syndulko K, Rever B, Kraut J, Coburn J, Tourtellotte WW. 1983. Visual evoked potentials and long latency event-related potentials in chronic renal failure. Neurology 33: 1219–22.

22. Collins DWK, Carroll WM, Black JL, Walsh M. 1979. Effect of refractive errors on the visual evoked response. Br Med J 1:231–32.

23. Conrad B, Banacke R, Musers H, Prange H, Behrens-Baumann W. 1983. Visual evoked potentials in neurosyphilis. J Neurol Neurosurg Psychiatry 46:23–27.

24. Cox C, Hack M, Metz D. 1981. Brainstem-evoked response audiometry: Normative data from the preterm infant. Audiology 20:53–64.

25. Desmedt JE, Cheron G. 1980. Somatosensory evoked potentials to finger stimulation in healthy octogenarians and in young adults: Wave forms, scalp topography and transit times of parietal and frontal components. Electroencephalogr Clin Neurophysiol 50:404–425.

26. DeVries LS, Pierrat V, Eken P, Minami T, Daniels H, Casaer P. 1991. Prognostic value of early somatosensory evoked potentials for adverse outcome in full-term infants with birth asphyxia. Brain Dev 13:320–25.

27. Dorfman LJ, Bosley TM. 1979. Age-related changes in peripheral and central nerve conduction in man. Neurology 29:38–44.

28. Drake ME, Pakalnis A, Hietter SA, Padamadan H. 1990. Visual and auditory evoked potentials in migraine. EMG Clin Neurophysiol 30:77–81.

29. Drake ME Jr, Pakalnis A, Padamadan H, Hietter SA. 1990. Auditory evoked potentials in vertebrobasilar transient ischemic attacks. Clin EEG 21:96–100.

30. Dustman RE, Beck EC. 1969. The effects of maturation and aging on the wave form of visually evoked potentials. EEG Clin Neurophysiol 26:2–11.

31. Eggermont JJ. 1992. Developmet of auditory evoked potentials Acta Oto-Laryngologica 112:197–200.

32. Ehle AL, Steward RM, Lellelid NA, Leventhal NA. 1984. Evoked potentials in Huntington's disease: A comparative and longitudinal study. Arch Neurol 41:379–82.

33. Fava E, Bortolani E, Ducati A, Schieppati M. 1992. Role of SEP in identifying patients requiring temporary shunt during carotid endarterectomy. EEG Clin Neurophysiol 84:426–32.

34. Febert A, Buchner H, Bruckmann H. 1990. Brainstem auditory evoked potentials and somatosensory evoked potentials in pontine hemorrhage. Correlations with clinical and CT findings. Brain 113:49–63.

35. Fenwick PBC, Brown D, Hennessey J. 1981. The visual evoked response to pattern reversal in normal 6–11 year-old children. EEG Clin Neurophysiol 51:49–62.

36. Fischer C, Blanc A, Mauguière F, Courjon J. 1981. Diagnostic value of brainstem auditory evoked potentials. Rev Neurol 137:229–240.

37. Fukai M, Motomura N, Kobayashi S, Asaba H, Sakai T. 1990. Event-related potential (P300) in epilepsy. Acta Neurol Scand 82:197–202.

38. Gibson NA, Graham M, Levene MI. 1992. Somatosensory evoked potentials and outcome in perinatal asphyxia. Arch Dis Child 67:393–98.

39. Gronfors T, Juhola M, Johansson R. 1992. Evaluation of some nonrecursive digital filters for signals of auditory evoked responses. Biological Cybernetics 66:533–36.

40. Halliday AM, McDonald WI, Mushin J. 1972. Delayed visual evoked response in optic neuritis. Lancet 1:982–85.

41. Hammond EJ, Wilder BJ. 1983. Evoked potentials in olivopontocerebellar atrophy. Arch Neurol 40:366–69.

42. Haupt WF, Horsch S. 1992. Evoked potential monitoring in carotid surgery: A review of 994 cases. Neurology 42:835–38.

43. Jiang ZD, Liu XY, Wu YY, Zheng MS, Liu HC. 1990. Long-term impairments of brain and auditory functions of children recovered from purulent meningitis. Dev Med Child Neurol 32:473–80.

44. Kaplan PW, Tusa RJ, Shankroff J, Heller J, Moser HW. 1993. Visual evoked potentials in adrenoleukodystrophy: A trial with glycerol trioleate and Lorenzo Oil. Ann Neurol 34:169–74.

45. Karhu J, Hari R, Makela JP, Huttunen J, Knuutila J. 1992. Cortical somatosensory magnetic responses in multiple sclerosis. EEG Clin Neurophysiol 83:192–200.

46. Kelly JJ, Sharbrough FW, Daube JR. 1981. A clinical and electrophysiological evaluation of myoclonus. Neurology 31:581–89.

47. King PJL, Chiappa KH. 1990. Motor evoked potentials. In Evoked potentials in clinical medicine, ed. KH Chiappa, Raven, New York.

48. Kinney JAS, McKay CL, Mensch AJ, Luria SM. 1973. Visual evoked responses elicited by rapid stimulation. EEG Clin Neurophysiol 34:7–13.

49. Klein AJ. 1983. Properties of the brainstem response slow-wave component. I. Latency, amplitude, and threshold sensitivity. Arch Otolaryngol 109:6–12.

50. Klein AJ. 1983. Properties of the brainstem response slow-wave component. II. Frequency specificity. Arch Otolaryngol 109:74–78.

51. Krieger D, Adams HP, Albert F, von Haken M, Hacke W. 1992. Pure motor hemiparesis with stable somatosensory evoked potential monitoring during aneurysm surgery: Case report. Neurosurgery 31:145–50.

52. Lowitzsch K, Rudolph HD, Trincker D, Muller E. 1980. Flash and pattern-reversal visual evoked responses in retrobulbar-neuritis and controls: A comparison of conventional and TV stimulation techniques. In EEG and clinical neurophysiology, Proceedings of the 2nd European congress of EEG and clinical neurophysiology, Salzburg, Austria, September 16–19, 1979, Excerpta Medica International Congress Series

No. 526, ed. H Lechner, A Aranibar, pp. 451–63. Amsterdam: Elsevier North-Holland.

53. Marra TR. 1990. The clinical significance of the bifid or "W" pattern reversal visual evoked potential. Clin EEG 21:162–67.

54. Marshall NK, Donchin E. 1981. Circadian variations in the latency of brainstem responses and its relation to body temperature. Science 212:356–58.

55. May JG, McCullen JK, Moskowitz-Cook A, Siegfried JB. 1979. Effects of meridional variation on steady-state visual evoked potentials. Vision Res 19:1395–1401.

56. Misulis KE. 1993. Essentials of clinical neurophysiology. Butterworth-Heinemann, Boston.

57. Molitor H. 1993. Somatosensory evoked potentials in root lesions and stenosis of the spinal canal (their diagnostic significance in clinical decision making). Neurosurg Rev 16:39–44.

58. Moller AR, Moller MB, Janetta PJ, Jho HD. 1991. Auditory nerve compound action potentials and brain stem auditory evoked potentials in patients with various degrees of hearing loss. Ann Otol Rhino Laryngol 100:488–95.

59. Mukhopadhyay S, Dhamija RM, Selvamurthy W, Chaturvedi RC, Thakur L, Sapra ML. 1992. Auditory evoked response in patients of diabetes mellitus. Indian I Med Res 96:81–86.

60. Nakashima K, Nitta T, Takahashi K. 1992. Recovery functions of somatosensory evoked potentials in parkinsonian patients. J Neurol Sci 108:24–31.

61. Oepen G, Doerr M, Thoden U. 1982. Huntington's disease: Alterations of visual and somatosensory cortical evoked potentials in patients and offspring. Adv Neurol 32:141–47.

62. Ollo C, Johnson R Jr, Grafman J. 1991. Signs of cognitive change in HIV disease: An event-related brain potential study. Neurology 41:209–15.

63. Ozdamar O, Kraus N. 1983. Auditory middle-latency responses in humans. Audiology 22:34–49.

64. Pakalnis A, Drake ME, Huber S, Paulson G, Phillips B. 1992. Central conduction time in progressive supranuclear palsy. EMG Clin Neurophysiol 32:41–42.

65. Pakalnis A, Drake ME Jr, Barohn RJ, Chakeres DW, Mendell JR. 1988. Evoked potentials in chronic inflammatory demyelinating polyneuropathy. Arch Neurol 45:1014–16.

66. Paludetti G, Maurizi M, Ottaviani F, Rosignoli M. 1981. Reference values and characteristics of brain stem audiometry in neonates and children. Scand Audiol 10:177–86.

67. Phillips KR, Potvin JH. 1983. Multimodality evoked potentials and neurophysiological tests in multiple sclerosis: Effects of hyperthermia on test results. Arch Neurol 40:159–64.

68. Polich J, Ladish C, Bloom FE. 1990. P300 assessment of early Alzheimer's disease. EEG Clin Neurophysiol 77:179–89.

69. Robinson K, Rudge P. 1981. Wave form analysis of the brain stem auditory evoked potential. EEG Clin Neurophysiol 52:583–94.

70. Rossini PM, Caltagirone C, David P, Macchi G. 1979. Jakob-Creutzfeldt disease: Analysis of EEG and evoked potentials under basal conductions and neuroactive drugs. Eur Neurol 18:269–79.

71. Rothwell JC, Obesco JA, Marsden CD. 1984. On the significance of giant somatosensory evoked potentials in cortical myoclonus. J Neurol Neurosurg Psychiatry 47:533–42.

72. Rowe MJ. 1978. Normal variability of the brainstem auditory evoked response in young and old adult subjects. EEG Clin Neurophysiol 44:459–70.

73. Salas-Puig J, Tunon A, Diaz M, Lahoz CH. 1992. Somatosensory evoked potentials in juvenile myoclonic epilepsy. Epilepsia 33:527–30.

74. Sato T, Miyao M, Muchi H, Gunji Y, Iizuka A, Yanagisawa M. 1992. P300 as indicator of prophylactic cranial radiation. Ped Neurol 8:130–32.

75. Shaw NA, Cant BR. 1981. Age-dependent changes in the amplitude of the pattern visual evoked potential. EEG Clin Neurophysiol 51:671–73.

76. Shaw NA, Cant BR. 1980. Age-dependent changes in the latency of the pattern visual evoked potential. EEG Clin Neurophysiol 48:237–41.

77. Shibasaki H, Yamashita Y, Kuroiwa Y. 1978. Electroencephalographic studies of myoclonus, myoclonus-related cortical spikes and high-amplitude somatosensory evoked potentials. Brain 101:447–60.

78. Sininger YS, Masuda A. 1990. Effect of click polarity on ABR threshold. Ear & Hearing 11:206–209.

79. Sokol S, Moskowitz A, Paul A. 1983. Evoked potential estimates of visual accommodation in infants. Vision Res 23:851–60.

80. Spehlmann R. 1965. The averaged electrical responses to diffuse and to patterned light in the human. EEG Clin Neurophysiol 19:560–69.

81. Stillman RD, Crow G, Moushegian G. 1978. Components of the frequency-following potential in man. EEG Clin Neurophysiol 44:438–46.

82. Stockard JJ, Hughes JF, Sharbrough FW. 1979. Visually evoked potentials to electronic pattern reversal: Latency variations with gender, age, and technical factors. Am J EEG Tech 19:171–204.

83. Stockard JJ, Pope-Stockard JE, Sharbrough FW. 1992. Brainstem auditory evoked potentials in neurology: Methodology, interpretation, and clinical application. In Electrodiagnosis in clinical neurology, ed. MJ Aminoff, pp. 503–536, Churchill Livingstone, New York.

84. Tandon OP, Misra R, Tandon I. 1990. Brainstem auditory evoked potentials (BAEPs) in pregnant women. Indian J Physiol Pharmacol 34:42–44.

85. Thivierge J, Cote R. 1990. Brainstem auditory evoked response: Normative values in children. EEG Clin Neurophysiol 77:309–13.

86. Towle VL, Maselli R, Bernstein LP, Spire JP. 1989. Electrophysiologic studies on locked-in patients: Heterogeneity of findings. EEG Clin Neurophysiol 73:419–26.

87. Uncini A, Treviso M, Basciani M, Onofrj M, Gambi D. 1988. Associated central and peripheral demyelination: An electrophysiological study. J Neurol 235:238–40.

88. Van Lith GHM, Van der Torren K, Vijfvinkel-Bruinenga S. 1983. Cataract, pattern stimulation and visually evoked potentials. Doc Ophthalmol 50:291–97.

89. Walk D, Fisher MA, Doundoulakis SH, Hemmati M. 1992. Somatosensory evoked potentials in the evaluation of lumbosacral radiculopathy. Neurology 42:1197–202.

90. Westheimer G. 1966. The Maxwellian view. Vision Res 6:669–82.

91. Williamson JB, Galasko CS. 1992. Spinal cord monitoring during operative correction of neuromuscular scoliosis. J Bone Joint Surg Br 74:w0-2.

92. Yamada O, Yamane H, Kodera K. 1977. Simultaneous recordings of the brainstem response and the frequency-following response to low-frequency tone. EEG Clin Neurophysiol 43:362–70.

93. Yoshie N. 1973. Diagnostic significance of the electrocochleogram in clinical audiometry. Audiology 12:504–39.

# Index